6
Modern
Mathematics
for Schools

6

Modern Mathematics for Schools

Second Edition

Scottish Mathematics Group

Blackie

Chambers

Blackie & Son Limited
Bishopbriggs · Glasgow G64 2NZ
7 Leicester Place
London WC2H 7BP

W & R Chambers Limited
43–45 Annandale Street · Edinburgh EH7 4AZ

© *Scottish Mathematics Group 1973*
First Published 1973
This reprint 1989

Designed by James W. Murray

International Standard Book Numbers
Pupils' Book
Blackie 0 216 89420 4
Chambers 0 550 75917 4
Teachers' Book
Blackie 0 216 89421 2
Chambers 0 550 75927 1

Printed in Great Britain by
Thomson Litho Ltd, East Kilbride, Scotland
Set in 10pt Monotype Times Roman

Members associated with this book

W. T. Blackburn
Formerly of Dundee College of Education
W. Brodie
Trinity Academy
C. Clark
Formerly of Lenzie Academy
D. Donald
Formerly of Robert Gordon's College
R. A. Finlayson
Formerly of Allan Glen's School
Elizabeth K. Henderson
Westbourne School for Girls
J. L. Hodge
Madras College
J. Hunter
University of Glasgow
R. McKendrick
Langside College
W. More
Formerly of High School of Dundee
Helen C. Murdoch
Hutchesons' Girls' Grammar School
A. G. Robertson
John Neilson High School
A. G. Sillitto
Formerly of Jordanhill College of Education
A. A. Sturrock
Formerly of Grove Academy
Rev. J. Taylor
Formerly of St. Aloysius' College
E. B. C. Thornton
Formerly of Bishop Otter College
J. A. Walker
Dollar Academy
P. Whyte
Hutchesons' Boys' Grammar School
H. S. Wylie
Formerly of Govan High School

Contributor of the Computer Studies
A. W. McMeeken
Dundee College of Education

Preface

Book 1 of the original series *Modern Mathematics for Schools* was first published in July 1965. This revised series has been produced in order to take advantage of the experience gained in the classroom with the original textbooks and to reflect the changing mathematical needs in recent years, particularly as a result of the general move towards some form of comprehensive education.

Throughout the whole series, the text and exercises have been cut or augmented wherever this was considered to be necessary, and nearly every chapter has been completely rewritten. In order to cater more adequately for the wider range of pupils now taking certificate-oriented courses, the pace has been slowed down in the earlier books in particular, and parallel sets of A and B exercises have been introduced where appropriate. The A sets are easier than the B sets, and provide straightforward but comprehensive practice; the B sets have been designed for the more able pupils, and may be taken in addition to, or instead of, the A sets. Often from Book 4 onwards a basic exercise, which should be taken by all pupils, is followed by a harder one on the same work in order to give abler pupils an extra challenge, or further practice; in such a case the numbering is, for example, Exercise 2 followed by Exercise 2B. It is hoped that this arrangement, along with the *Graph Workbook for Modern Mathematics*, will allow considerable flexibility of use, so that while all the pupils in a class may be studying the same topic, each pupil may be working examples which are appropriate to his or her aptitude and ability.

In 1974 the first of a series of pads of expendable *Mathsheets* was published in order to provide simpler and more practical material; each pad contains a worksheet for every exercise in the corresponding textbook. This increases still further the flexibility of the *Modern Mathematics for Schools* 'package' by providing three closely related levels of work—mathsheets, A exercises and B exercises. *Mathsheets* to accompany Books 1, 2 and 3 are now available.

Each chapter is backed up by a summary, and by revision exercises; in addition, cumulative revision exercises have been introduced at the end of alternate books. A new feature is the series of Computer Topics from Book 4 onwards. These form an elementary introduction to computer studies, and are primarily intended to give pupils some appreciation of the applications and influence of computers in modern society.

Books 1 to 7 provide a suitable course for many modern Ordinary Level and Ordinary Grade syllabuses in mathematics, including the University of London GCE Syllabus C, the Associated Examining Board Syllabus Mathematics 105, the Cambridge Local Syndicate Syllabus C, and the Scottish Certificate of Education. Books 8 and 9 complete the work for the Scottish Higher Grade Syllabus, and provide a good preparation for all Advanced Level and Sixth Year Syllabuses, both new and traditional.

Related to this revised series of textbooks are the *Modern Mathematics Newsletters*, the *Teacher's Editions* of the textbooks, the *Graph Workbook for Modern Mathematics*, the *Three-Figure Tables for Modern Mathematics*, and the booklets of *Progress Papers for Modern Mathematics*. These new Progress Papers consist of short, quickly marked objective tests closely connected with the textbooks. There is one booklet for each textbook, containing A and B tests on each chapter, so that teachers can readily assess their pupils' attainments, and pupils can be encouraged in their progress through the course.

The separate headings of Algebra, Geometry, Arithmetic, and later Trigonometry and Calculus, have been retained in order to allow teachers to develop the course in the way they consider best. Throughout, however, ideas, material and method are integrated *within* each branch of mathematics and *across* the branches; the opportunity to do this is indeed one of the more obvious reasons for teaching this kind of mathematics in the schools—for it is *mathematics* as a whole that is presented.

Pupils are encouraged to find out facts and discover results for themselves, to observe and study the themes and patterns that pervade mathematics today. As a course based on this series of books progresses, a certain amount of equipment will be helpful, particularly in the development of geometry. The use of calculating machines, slide rules, and computers is advocated where appropriate, but these instruments are not an essential feature of the work.

While fundamental principles are emphasised, and reasonable attention is paid to the matter of structure, the width of the course should be sufficient to provide a useful experience of mathematics for those pupils who do not pursue the study of the subject beyond school level. An effort has been made throughout to arouse the interest of all pupils and at the same time to keep in mind the needs of the future mathematician.

The introduction of mathematics in the Primary School and recent changes in courses at Colleges and Universities have been taken into account. In addition, the aims, methods, and writing of these books have been influenced by national and international discussions about the purpose and content of courses in mathematics, held under the auspices of the Organisation for Economic Co-operation and Development and other organisations.

The authors wish to express their gratitude to the many teachers who have offered suggestions and criticisms concerning the original series of textbooks; they are confident that as a result of these contacts the new series will be more useful than it would otherwise have been.

Contents

Algebra

Geometry

Arithmetic

Trigonometry

Computer Studies

Notation

Sets of numbers

Different countries and different authors
give different notations and definitions
for the various sets of numbers.
In this series the following are used:

E The universal set

$\emptyset$ The empty set

N The set of natural numbers $\{1, 2, 3, ...\}$

W The set of whole numbers $\{0, 1, 2, 3, ...\}$

Z The set of integers $\{..., -2, -1, 0, 1, 2, ...\}$

Q The set of rational numbers

R The set of real numbers

The set of prime numbers $\{2, 3, 5, 7, 11, ...\}$

Algebra

Factors and Fractions

1 The distributive law and common factors

The distributive law can be stated as follows:

$$ab + ac = a(b+c),$$

for all a, b, $c \in R$.

It shows how to express a sum of terms, which have a factor in common, as a product. Thus, the factor a in each term on the left side can be 'taken out' as a *common factor* of the whole expression, as shown on the right side.

Just as in $15 = 3 \times 5$,   3 and 5 are factors of 15, so in

$ab + ac = a(b+c)$, a and $(b+c)$ are factors of $ab+ac$.

Just as we divide 15 by 3 to find the factor 5, so we divide each term of $ab+ac$ by a to find the factor $(b+c)$.

Example 1. Factorise $pq + p$.

The common factor is p, and dividing each term by p gives the other factor $q+1$.

So
$$pq + p$$
$$= p(q+1)$$

Example 2. Factorise $4x^2 - 6x^5$.

$$4x^2 - 6x^5$$
$$= 2x^2(2 - 3x^3)$$

Example 3. Express in factors $x^2yz + xy^2z + xyz^2$.

The highest common factor of the three terms is xyz.

$$x^2yz + xy^2z + xyz^2$$
$$= xyz(x + y + z)$$

Note that we can always check our answer by multiplying out, using the distributive law.

Exercise 1

Copy and complete:

1 $2a + 2b = 2(\quad)$ *2* $3x + 6y = 3(\quad)$

3 $6p + 4q = 2(\quad)$ *4* $ap + aq = a(\quad)$

5 $ab + a = a(\quad)$ *6* $xy + x^2 = x(\quad)$

7 $ac^2 + a = a(\quad)$ *8* $2ax + 4ay = 2a(\quad)$

9 $3c - 3d = 3(\quad)$ *10* $12m - 8n = 4(\quad)$

11 $4pq - 8p = 4p(\quad)$ *12* $6x - 15x^2 = 3x(\quad)$

Factorise the following by taking out the common factors:

13 $3a + 3b$ *14* $8x + 8y$ *15* $4x - 4y$ *16* $5c - 5d$

17 $2b + 4c$ *18* $9m - 6n$ *19* $ax + ay$ *20* $bx - by$

21 $ax + a$ *22* $cx^2 + c$ *23* $a^2 + ab$ *24* $p^2 - pq$

25 $x^2 + x$ *26* $y^2 - y$ *27* $t^3 + t^2$ *28* $k^3 - k^2$

29 $pq + qr$ *30* $cd - de$ *31* $8x + 12y$ *32* $35a - 14b$

33 $2a^2 + 6ab$ *34* $12ac - 9ab$ *35* $abc + abd$ *36* $6n^2 - 2n$

37 The use of factors can sometimes simplify calculation. Find a quick way to calculate the following:

a $(34 \times 57) + (34 \times 43)$ *b* $(5{\cdot}4 \times 38) - (5{\cdot}4 \times 36)$

c $(3{\cdot}4 \times 7{\cdot}8) + (2{\cdot}2 \times 3{\cdot}4)$ *d* $(2{\cdot}5 \times 6{\cdot}5) - (2{\cdot}5 \times 2{\cdot}5)$

Express in factorised form questions *38* to *41*:

38 $2a + 2b + 2c$ *39* $ax + ay + az$

40 $cx - cy + cz$ *41* $4x + 2y - 6z$

42 Express the formula $Q = 4p^2 - 10pq$ in a fully factorised form. Calculate Q when $p = 9{\cdot}5$ and $q = 3$.

Exercise 1B

Factorise:

1	$6m+9n$	*2*	$12p+8q$	*3*	$ax-bx$	*4*	$20-4x$	
5	$x-x^2$	*6*	$xy+y$	*7*	$mv-mu$	*8*	$2ax+ay$	
9	$6ax-4ay$	*10*	$9ab-12ac$	*11*	$6x^2+4x$	*12*	$15a^2-6ab$	
13	$\frac{1}{3}p-\frac{1}{3}q$	*14*	$\frac{1}{2}ah+\frac{1}{2}bh$	*15*	$ut+\frac{1}{2}at^2$	*16*	$10p^2q+8pq^2$	
17	$\frac{1}{2}gtT-\frac{1}{2}gt^2$	*18*	$\frac{1}{2}+\frac{1}{2}a$	*19*	$\frac{1}{2}m-\frac{3}{2}$	*20*	$2\pi rh+2\pi r^2$	
21	$3a+6b+9c$	*22*	$4x+2y+8z$		*23*	$ab+ac-a^2$		
24	$9p-6q-3r$	*25*	$3x^2-2xy+3x$		*26*	$15x^2+10xy-20xz$		

Which of the following are true and which are false?

27 $p^2+pq+p = p(p+q)$ *28* $15-3x = 5(3-x)$

29 $a\sin A+b\sin A = \sin A\,(a+b)$ *30* $\tan^2 A+\tan A = \tan A(\tan A+1)$

31 $2L+2B = 4LB$ *32* $\pi r^2+2\pi rh = \pi r(r+2h)$

33 $t^2+t^4+t^6 = t^2(1+t^2+t^3)$ *34* $\frac{2}{3}\pi r^3+\pi r^2 h = \frac{2}{3}\pi r^2(r+h)$

35 $\dfrac{1}{x}+\dfrac{1}{x^2} = \dfrac{1}{x}\left(1+\dfrac{1}{x}\right)$ *36* $\dfrac{2}{a}+\dfrac{4}{a^2}+\dfrac{6}{a^3} = \dfrac{2}{a}\left(1+\dfrac{2}{a}+\dfrac{3}{a^2}\right)$

Copy and complete:

37 $a-b+c = a-(\qquad)$ *38* $a-b-c = a-(\qquad)$

39 $a-3b-6c = a-3(\qquad)$ *40* $a-3b+6c = a-3(\qquad)$

41 $a^2+ax+ay+xy = a(\qquad)+y(\qquad)$

42 $ab-ax-bx+x^2 = a(\qquad)-x(\qquad)$

43 $(a+b)x+(a+b)y = (a+b)(\qquad)$

44 $a(a+x)+b(a+x) = (a+x)(\qquad)$

45 $a(a+b)-c(a+b) = (a+b)(\qquad)$

46 $(c-a)x-y(c-a) = (c-a)(\qquad)$

By taking out a common factor of the first two terms and then of the second two terms, express each of the following as a product of two factors:

47 $a^2 + ab + ac + bc$ *48* $ax - ay + bx - by$

49 $a^2 + ab - ac - bc$ *50* $a^2 - ab - ac + bc$

---

2 *Difference of two squares*

If a and b are variables on the set of real numbers, we have, using the distributive law,

$$(a-b)(a+b) = a(a+b) - b(a+b)$$
$$= a^2 + ab - ba - b^2$$
$$= a^2 + ab - ab - b^2$$
$$= a^2 - b^2$$

It follows that:

$$a^2 - b^2 = (a-b)(a+b)$$

On the left-hand side we have a *difference of two squares*, and on the right-hand side a *product of two factors*.

Example 1. Factorise $a^2 - 16$ *Example 2*. Factorise $2x^2 - 18y^2$

$$a^2 - 16 \qquad\qquad\qquad 2x^2 - 18y^2$$
$$= a^2 - 4^2 \qquad\qquad\quad = 2(x^2 - 9y^2)$$
$$= (a-4)(a+4) \qquad = 2[x^2 - (3y)^2]$$
$$\qquad\qquad\qquad\qquad = 2(x-3y)(x+3y)$$

Note that any common factors must be taken out first.

Exercise 2

Factorise the following, all of which contain a difference of two squares:

1 $a^2 - b^2$ *2* $c^2 - d^2$ *3* $p^2 - q^2$ *4* $x^2 - y^2$

5 $x^2 - 2^2$ *6* $a^2 - 3^2$ *7* $b^2 - 4^2$ *8* $c^2 - 5^2$

9	a^2-1	10	b^2-4	11	c^2-36	12	d^2-100
13	$1-k^2$	14	$9-m^2$	15	$16-n^2$	16	$64-p^2$
17	r^2-s^2	18	u^2-v^2	19	m^2-n^2	20	y^2-z^2
21	a^2-4b^2	22	c^2-9d^2	23	e^2-49f^2	24	g^2-81h^2
25	$16x^2-y^2$	26	$4p^2-q^2$	27	$25r^2-s^2$	28	$36u^2-v^2$
29	$4x^2-9y^2$	30	$9a^2-16b^2$	31	$16c^2-9d^2$	32	$4e^2-25f^2$
33	$2a^2-2b^2$	34	$2a^2-8b^2$	35	$2a^2-18b^2$	36	$2a^2-50b^2$
37	$5p^2-5q^2$	38	$45p^2-5q^2$	39	$3p^2-12q^2$	40	$27p^2-3q^2$
41	$9a^2-36y^2$	42	$3b^2-27$	43	$4c^2-100$	44	$5d^2-5$
45	$8x^2-32y^2$	46	$8x^2-50y^2$	47	$p^2-4q^2r^2$	48	$a^2b^2-4c^2$

Example 3. Factorise k^4-1

$$k^4-1$$
$$=(k^2)^2-1^2$$
$$=(k^2-1)(k^2+1)$$
$$=(k-1)(k+1)(k^2+1)$$

Example 4. Factorise $a^2-(b-c)^2$

$$a^2-(b-c)^2$$
$$=[a-(b-c)][a+(b-c)]$$
$$=(a-b+c)(a+b-c)$$

Exercise 2B

Factorise completely:

1	a^4-1	2	c^4-16	3	d^4-81	4	$1-x^4$
5	a^4-b^4	6	x^4-y^4	7	p^4-q^4	8	$1-16c^4$
9	$3x^4-48$	10	$4y^4-4$	11	$2z^4-162$	12	$80-5a^4$

13 $(a+b)^2-c^2$ 14 $(x+y)^2-z^2$ 15 $(p+q)^2-4$

16 $(m-n)^2-p^2$ 17 $(b-c)^2-a^2$ 18 $(m-n)^2-1$

19 $a^2-(b-c)^2$ 20 $d^2-(e-f)^2$ 21 $4f^2-(g-h)^2$

22 $1-(x-y)^2$ 23 $4-(m-n)^2$ 24 $9-(a+b)^2$

25 $(a+b)^2-(a-b)^2$ 26 $(p-q)^2-(p+q)^2$

27 $16x^2-9(x-y)^2$ 28 $25a^2-4(a+b)^2$

Some applications of the difference of two squares

We can illustrate the identity $a^2 - b^2 = (a-b)(a+b)$ as in Figure 1.

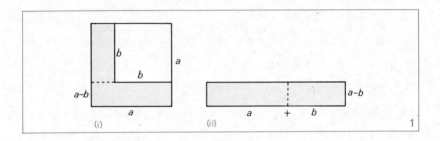

(i)   (ii)   1

The area of the larger square is a^2, and the area of the smaller square is b^2, so that the area of the shaded L-shape is $a^2 - b^2$.

If the squares are cut out in cardboard, and the L-shape is cut along the dotted line, the two shaded pieces can be put together as in the second diagram to form a rectangle with area $(a-b)(a+b)$. Hence $a^2 - b^2 = (a-b)(a+b)$.

Example 1. Use factors to calculate $7{\cdot}7^2 - 2{\cdot}3^2$

$$7{\cdot}7^2 - 2{\cdot}3^2$$
$$= (7{\cdot}7 - 2{\cdot}3)(7{\cdot}7 + 2{\cdot}3)$$
$$= 5{\cdot}4 \times 10$$
$$= 54$$

Example 2. Triangle ABC is right-angled at A. Calculate b when $a = 58$, $c = 42$.

By Pythagoras' theorem,
$$a^2 = b^2 + c^2.$$
So $\quad b^2 = a^2 - c^2$
$$= 58^2 - 42^2$$
$$= (58 - 42)(58 + 42)$$
$$= 16 \times 100$$
$$= 1600$$

Hence $\quad b = 40$

Exercise 3

Use factors to calculate the values in questions *1–8*:

1 $\quad 9^2 - 7^2$ $\qquad$ *2* $\quad 11^2 - 5^2$ $\qquad$ *3* $\quad 20^2 - 19^2$ $\qquad$ *4* $\quad 63^2 - 37^2$

5 $\quad 9{\cdot}5^2 - 7{\cdot}5^2$ $\quad$ *6* $\quad 7{\cdot}3^2 - 2{\cdot}7^2$ $\quad$ *7* $\quad 99^2 - 1^2$ $\quad$ *8* $\quad 999^2 - 1$

In questions *9–12*, a, b, c are the sides of $\triangle$ABC which is right-angled at A. It may be helpful to draw a sketch.

9 $a = 25,\ c = 24$; find b *10* $a = 41,\ c = 40$; find b

11 $a = 37,\ b = 35$; find c *12* $a = 61,\ b = 60$; find c

13 Figure 2 shows a square metal plate of side X cm with a square portion of side x cm removed. Write down a formula for the area A cm^2 of metal remaining. Calculate A when:

a $X = 7\cdot5$ and $x = 2\cdot5$ *b* $X = 25$ and $x = 15$

c $X = 6\cdot5$ and $x = 2\cdot5$ *d* $X = 0\cdot64$ and $x = 0\cdot24$

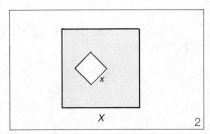
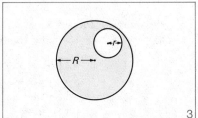

14 Factorise completely $\pi R^2 - \pi r^2$. *Hence* calculate the shaded area between the two circles in Figure 3 in the cases where

a $R = 15$ and $r = 5$ *b* $R = 8\cdot5$ and $r = 1\cdot5$

(Take $3\cdot14$ as an approximation for π; the units are centimetres.)

15 A ventilator consists of a circular metal plate of radius R cm with 100 separate holes drilled in it, each of radius r cm. Find a formula for the area of metal in the ventilator.

  Calculate the area when $R = 35$ and $r = 1\cdot5$. (Take $3\cdot14$ as an approximation for π.)

16 The expression $\frac{1}{2}mv^2 - \frac{1}{2}mu^2$ occurs in mechanics. Write this in a fully factorised form, and hence calculate its value when $m = 10$, $v = 74$ and $u = 26$.

17 Factorise the following trigonometrical expressions, remembering that $\cos^2 x$ means $(\cos x)^2$.

a $\cos^2 x - \sin^2 x$ *b* $1 - \tan^2 x$

c $4\sin^2 x - 1$ *d* $9 - 4\cos^2 x$

18 A metal cuboid has length l cm and has a square cross-section of side a cm. Four separate holes with square cross-section of side b cm ($b<\frac{1}{2}a$) are cut through the metal, parallel to the length.

a Show that the volume, V cm^3, of the metal left is given by $V = l(a-2b)(a+2b)$.

b Calculate V when $a = 15$, $b = 4.5$, $l = 20$.

---

3 Quadratic expressions and their factors

In Algebra, Book 5, Chapter 1, we obtained quadratic expressions like x^2+5x+6 by using the distributive law as follows:

$$
\begin{array}{ll}
\text{(i)} & (x+2)(x+3) \\
& = x(x+3)+2(x+3) \\
& = x^2+3x+2x+6 \\
& = x^2+5x+6
\end{array}
\qquad
\begin{array}{ll}
\text{(ii)} & (x-3)(x-5) \\
& = x(x-5)-3(x-5) \\
& = x^2-5x-3x+15 \\
& = x^2-8x+15
\end{array}
$$

In this chapter we study ways of factorising *quadratic expressions* (or *trinomials*, or *polynomials of degree* 2) of the form ax^2+bx+c, and we shall use this new skill when we solve quadratic equations in Chapter 3 of the Algebra in this book.

Part 1: ax^2+bx+c, $a=1$ and $c>0$.

From the above examples,

$x^2+5x+6 = (x+2)(x+3)$. Note that $6 = 2\times3$, and $5 = 2+3$.

$x^2-8x+15=(x-3)(x-5)$. Note that $15 = -3\times(-5)$,
and $-8 = -3+(-5)$.

In each case:

(i)   the *constant* in the quadratic expression is equal to the *product* of the constants in the brackets (and note that as the sign of the constant in the quadratic expression is $+$, the signs of the constants in the brackets must both be $+$, or both $-$)

(ii) the *coefficient of* x in the quadratic expression is equal to the *sum* of the constants in the brackets.

We can see the above clearly from the statement

$$x^2+(p+q)x+pq = (x+p)(x+q)$$

Example. Factorise $x^2 - 8x + 12$.

We must first find two numbers with product 12 and sum -8. These are -6 and -2, so
$$x^2 - 8x + 12 = (x - 6)(x - 2).$$

Exercise 4

Multiply out the following and in each case notice the relations given in (i) and (ii) above:

1 $(x + 3)(x + 4)$ *2* $(x + 1)(x + 7)$

3 $(x - 5)(x - 2)$ *4* $(x - 6)(x - 6)$

· In each of the following, give two numbers whose product is the first number and whose sum is the second number:

5   6, 5   *6*   20, 9   *7*   7, 8   *8*   9, 6   *9*   10, 11

10   10, -7 *11*   15, -8 *12*   30, -11 *13*   2, -3 *14*   8, -9

Copy and complete the following:

15 $x^2 + 7x + 12 = (x + 4)(x + \quad)$ *16* $x^2 + 6x + 5 = (x + \quad)(x + \quad)$

17 $x^2 - 6x + 8 = (x - 4)(x - \quad)$ *18* $x^2 - 3x + 2 = (x - \quad)(x - \quad)$

19 $x^2 + 8x + 15 = (x + 3)(\quad)$ *20* $x^2 - 12x + 20 = (x - 2)(\quad)$

Factorise:

21 $x^2 + 3x + 2$ *22* $a^2 - 3a + 2$ *23* $x^2 + 6x + 8$ *24* $p^2 - 6p + 8$

25 $x^2 + 9x + 18$ *26* $q^2 - 9q + 18$ *27* $k^2 - 8k + 12$ *28* $a^2 + 8a + 12$

29 $x^2 + 11x + 10$ *30* $z^2 + 10z + 21$ *31* $c^2 - 9c + 8$ *32* $n^2 + 12n + 32$

33 $x^2 + 11x + 24$ *34* $v^2 - 14v + 24$ *35* $b^2 + 10b + 25$ *36* $m^2 + 11m + 28$

37 $x^2 - 14x + 45$ *38* $d^2 + 18d + 45$ *39* $x^2 - 19x + 60$ *40* $y^2 + 24y + 80$

41 $6 + 5x + x^2$ *42* $15 - 8x + x^2$ *43* $50 + 15x + x^2$ *44* $36 - 20x + x^2$

45 $16 + 8x + x^2$ *46* $25 - 10a + a^2$ *47* $1 - 2b + b^2$ *48* $24 + 11x + x^2$

Part 2: $ax^2 + bx + c$, $a = 1$ and $c < 0$

Since
$$(x+5)(x-2) = x^2 + 3x - 10,$$
$$x^2 + 3x - 10 = (x+5)(x-2).$$

Note that $-10 = 5 \times (-2)$, and $3 = 5 + (-2)$.

Since
$$(x-4)(x+1) = x^2 - 3x - 4,$$
$$x^2 - 3x - 4 = (x-4)(x+1).$$

Note that $-4 = -4 \times 1$, and $-3 = -4 + 1$.

The same relations are true as in Part 1 above, but notice that since the sign of the constant in the quadratic expression is $-$, the signs of the constants in the brackets must be opposites. The sign of the coefficient of x in the quadratic expression is the sign of the numerically larger constant in the brackets.

Example. Factorise $x^2 - 2x - 24$.

Here we must first find two numbers whose product is -24 and whose sum is -2.

We have $6 \times 4 = 24$, $-6 \times 4 = -24$, $-6 + 4 = -2$; so the numbers are -6 and 4.

$$\text{Then } x^2 - 2x - 24 = (x-6)(x+4).$$

Exercise 5

Multiply out the following, and in each case notice the relations between the numbers in the brackets and the numbers in the quadratic expression:

1 $(x+6)(x-3)$ *2* $(x+8)(x-1)$

3 $(x-5)(x+2)$ *4* $(x-10)(x+1)$

In each of the following give two numbers whose product is the first number and whose sum is the second number:

5 $-20, 16$ $-7, 6$ *7* $-16, 0$ *8* $-9, 0$ *9* $-5, 4$

10 $-10, 3$ *11* $-18, -3$ *12* $-24, -5$ *13* $-36, -16$ *14* $-2, -1$

Copy and complete the following:

15 $x^2 + 4x - 12 = (x+6)(x\ \ )$ 16 $x^2 - 8x - 9 = (x+1)(x\ \ )$

17 $x^2 - x - 2 = (x-2)(\ \ )$ 18 $x^2 - 9x - 36 = (x-12)(\ \ )$

Factorise:

19 $a^2 + 3a - 10$ 20 $b^2 - 3b - 10$ 21 $c^2 + 6c - 7$ 22 $d^2 - 6d - 7$

23 $x^2 + 5x - 36$ 24 $x^2 - 5x - 36$ 25 $x^2 + 3x - 18$ 26 $x^2 - 3x - 18$

27 $m^2 - 2m - 24$ 28 $n^2 - 5n - 24$ 29 $p^2 - 7p - 30$ 30 $q^2 - 4q - 45$

31 $a^2 + a - 2$ 32 $b^2 + b - 6$ 33 $c^2 - c - 12$ 34 $d^2 - d - 20$

35 $p^2 + 2p - 15$ 36 $q^2 + 4q - 21$ 37 $r^2 + 6r - 7$ 38 $s^2 + 4s - 12$

39 $t^2 + 2t - 8$ 40 $u^2 - 2u - 8$ 41 $v^2 + 7v - 8$ 42 $w^2 - 7w - 8$

43 $x^2 + 10x - 24$ 44 $x^2 - x - 42$ 45 $a^2 + 2a - 48$ 46 $a^2 - 6a - 72$

47 $y^2 - y - 72$ 48 $y^2 - 21y - 72$ 49 $z^2 + 14z - 72$ 50 $z^2 + 71z - 72$

Part 3: $ax^2 + bx + c$, $a \neq 1$

We have $(2x+3)(4x-5) = 2x(4x-5) + 3(4x-5)$

$$= 2.4x^2 - 2.5x + 3.4x - 3.5$$
$$= 2.4x^2 + (3.4 - 2.5)x - 3.5$$
$$= 8x^2 + 2x - 15$$

Therefore $8x^2 + 2x - 15$, or $2.4x^2 + (3.4 - 2.5)x - 3.5$
$$= (2x+3)(4x-5).$$

Note that the factors can be written down by considering:
(i) the x-term in each bracket, so that their product is $8x^2$
(ii) the factors of the constant term, -15, so that the sum of the 'inner product' $(3 \times 4x)$ and the 'outer product' $(-5 \times 2x)$ is $2x$.

Example. Factorise $2x^2 - 9x - 18$.

$$2x^2 - 9x - 18 = (2x \quad)(x \quad)$$

inner product / outer product

(i) The factors of $2x^2$ are $2x$ and x.
(ii) The factors of 18 are 18, 1; 9, 2; 6, 3. We try these factors, with

appropriate signs, in the brackets until we find that the sum of the inner product and the outer product is $-9x$.

Then $2x^2 - 9x - 18 = (2x + 3)(x - 6)$.

Exercise 6

Factorise the following; *check each answer mentally.*

1	$2x^2 + 5x + 3$	*2*	$2x^2 + 7x + 3$	*3*	$3x^2 + 7x + 2$
4	$6a^2 + 7a + 2$	*5*	$10x^2 + 17x + 3$	*6*	$3x^2 + 14x + 15$
7	$9c^2 + 6c + 1$	*8*	$10x^2 + 19x + 6$	*9*	$12y^2 - 8y + 1$
10	$2n^2 - 7n + 3$	*11*	$2z^2 - 5z + 2$	*12*	$2x^2 - 5x + 3$
13	$8b^2 - 14b + 5$	*14*	$6a^2 - 13a + 6$	*15*	$9c^2 - 24c + 16$
16	$2p^2 - 19p + 9$	*17*	$12y^2 - 23y + 10$	*18*	$10m^2 - 43m + 12$
19	$3z^2 - 2z - 8$	*20*	$4p^2 + 7p - 2$	*21*	$3n^2 - 5n - 2$
22	$3k^2 + k - 2$	*23*	$6n^2 - 5n - 6$	*24*	$5y^2 + 4y - 1$
25	$6x^2 + 17x - 3$	*26*	$2x^2 + x - 1$	*27*	$3x^2 - 2x - 1$
28	$5x^2 - 23x - 10$	*29*	$8x^2 + 2x - 3$	*30*	$8c^2 + 10c - 3$
31	$6z^2 - 11z + 3$	*32*	$12a^2 - 11a - 5$	*33*	$4x^2 + 12x + 9$
34	$24y^2 + 2y - 1$	*35*	$1 + 3x - 18x^2$	*36*	$15 - 7p - 2p^2$

Exercise 6B

Factorise the following; *check each answer mentally.*

1	$a^2 + 8a + 12$	*2*	$a^2 + 8ab + 12b^2$	*3*	$p^2 + 8pq + 12q^2$
4	$c^2 - 10c + 24$	*5*	$x^2 - 10xy + 24y^2$	*6*	$c^2 - 10cd + 24d^2$
7	$a^2 + 3ab + 2b^2$	*8*	$p^2 - pq - 2q^2$	*9*	$r^2 + rs - 6s^2$
10	$u^2 - 2uv + v^2$	*11*	$a^2 - 5ab - 14b^2$	*12*	$c^2 + 7cd - 30d^2$
13	$2x^2 + 11xy + 5y^2$	*14*	$2y^2 + yz - z^2$	*15*	$4a^2 - 7ab - 2b^2$
16	$2a^2 - 5ab - 3b^2$	*17*	$2a^2 - 5ab + 3b^2$	*18*	$2a^2 + 5ab - 3b^2$

19	$9x^2 + 6xy - 8y^2$	20	$6r^2 - 5rs - 6s^2$	21	$4x^2 - 17xy + 4y^2$
22	$12a^2 + 13ab - 4b^2$	23	$9x^2 + xy - 10y^2$	24	$8a^2 + 7ab - 15b^2$
25	$2\cos^2\theta + 3\cos\theta - 2$	26	$3\sin^2\theta - 4\sin\theta + 1$		
27	$5\tan^2\theta + 2\tan\theta - 3$	28	$9\sin^2\theta - 12\sin\theta + 4$		
29	$6\tan^2\theta + 7\tan\theta - 20$	30	$10\cos^2\theta - \cos\theta - 3$		

4 Miscellaneous factors

In Sections 1–3 we studied factors of the following kinds:

(i) Common factors, e.g. $ax + ay = a(x + y)$.

(ii) Difference of two squares, e.g. $a^2 - b^2 = (a - b)(a + b)$.

(iii) Quadratic expressions, e.g. $2x^2 - 3x - 5 = (2x - 5)(x + 1)$.

We now use these in sets of miscellaneous examples.

Exercise 7

Factorise:

1	$5x + 15y$	2	$x^2 - 25$	3	$a^2 + 6a + 9$	4	$x^2 - x$
5	$y^2 - y - 6$	6	$1 - p^2$	7	$a^2 + 4a + 4$	8	$ax - ap + ad$
9	$3x^2 - 12$	10	$p^2 - q^2$	11	$p^2 - 2p$	12	$p^2 - 2p + 1$
13	$a^2 - 1$	14	$a^2 - a$	15	$a^2 - a - 2$	16	$2x^2 - 18$
17	$2x^2 - 8x$	18	$x^3 - x^2$	19	$2k^2 + 3k - 5$	20	$k^2 - 6k + 9$
21	$9k^2 + 6k + 1$	22	$16 - x^2$	23	$18 - 2y^2$	24	$12z - 6z^2$
25	$9 - 4a^2$	26	$2p^2 - p - 1$	27	$6x^2 + 13x + 6$	28	$14a^2 + 21b^2$
29	$14a^2 - 56b^2$	30	$16x^2 - 8x + 1$	31	$4a^2b - 8ab^2$	32	$1 - 2x + x^2$
33	$ax + bx - 2x$	34	$2x^3 - 32x$	35	$6y^2 + 5y - 6$	36	$2x^2 + 4x + 2$
37	$3a^2 + 3a - 6$	38	$4b^2 + 14b - 8$	39	$5x - 20x^3$	40	$6x + 24x^3$
41	$km^2 - kn^2$	42	$2c^2 - 7c - 15$	43	$a^2 + a^3 + a^4$	44	$4x^2 - 11x + 6$

Exercise 7B

Factorise completely:

1	$x^4 - x^6$	*2*	$x^4 - 1$	*3*	$(a-b)^2 - x^2$
4	$a^2 - 10ab + 25b^2$	*5*	$m^4 - n^4$	*6*	$4a^2b^3 - 12a^3b^2$
7	$1 - (p-q)^2$	*8*	$12a^2 + 7ab - 12b^2$	*9*	$16 - 8k + k^2$
10	$x^2y^6 - x^6y^2$	*11*	$2x^3 + 2x^2 - 4x$	*12*	$3 - 3x - 36x^2$
13	$x^2 - (y+z)^2$	*14*	$(x-3)^2 - 9$	*15*	$6a^2 - ab - 15b^2$
16	$2\tan^2 A - 4\tan A$	*17*	$4\cos^2 A - \sin^2 A$	*18*	$2\sin^2 A + 5\sin A + 3$
19	$2 - 2(x-y)^2$	*20*	$a(x-y) - b(x-y)$	*21*	$(p-q)^2 - (r-s)^2$
22	$x^5 - 81x$	*23*	$(x-y)^2 - 3(x-y)$	*24*	$a(p-q) - p + q$
25	$(x-1)^2 + 2(x-1)$	*26*	$(x+1)^3 - 4(x+1)$	*27*	$2(a-b) - 18(a-b)^3$
28	$12a^2 + 8a - 15$	*29*	$1 - x^8$	*30*	$6a^3b^4 - 9a^5b^3 + 3a^4b^5$
31	$r(r+1) - 3r$	*32*	$(r^2+r)^2 - (r^2-r)^2$	*33*	$n^2(n+1)^2 - n(n+1)$

---

5 Using factors to simplify fractional expressions

We can simplify $\dfrac{14}{20}$ by factorising 14 and 20 as follows:

$$\frac{14}{20} = \frac{2 \times 7}{2 \times 10} = \frac{7}{10}$$

In the same way we can often simplify fractional expressions by first factorising the numerator and denominator. A fractional expression is said to be *simplified*, or to be in its *lowest terms*, when its numerator and denominator have no common factor other than unity.

Example 1. Simplify $\dfrac{3x+12y}{6}$

$$\dfrac{3x+12y}{6} = \dfrac{3(x+4y)}{6}$$

$$= \dfrac{x+4y}{2}$$

Example 2. Reduce $\dfrac{2x-6}{x^2+x-12}$

to its lowest terms.

$$\dfrac{2x-6}{x^2+x-12} = \dfrac{2(x-3)}{(x+4)(x-3)}$$

$$= \dfrac{2}{x+4}$$

Note

(i) The entire numerator and denominator must be factorised before simplification.

(ii) In Example 2, we assume that $x \neq -4$ and $x \neq 3$ as in these cases the fraction would have no meaning. Throughout this section, it will be assumed that all denominators are non-zero.

Exercise 8

Simplify the following:

1 $\dfrac{18}{30}$ *2* $\dfrac{21}{45}$ *3* $\dfrac{22}{44}$ *4* $\dfrac{19}{57}$ *5* $\dfrac{24}{36}$

6 $\dfrac{4x+6}{2}$ *7* $\dfrac{12a+8}{4}$ *8* $\dfrac{14a+10b}{2}$ *9* $\dfrac{10p-20q}{5}$

10 $\dfrac{8x-16y}{2}$ *11* $\dfrac{9x-6y}{3}$ *12* $\dfrac{ab+ac}{a}$ *13* $\dfrac{a^2-3ab}{a}$

14 $\dfrac{12ab+3b^2}{3b}$ *15* $\dfrac{18a^2+12a^3}{6a}$ *16* $\dfrac{3a+3b}{6c}$ *17* $\dfrac{xy-zx}{2x}$

18 $\dfrac{3}{6a+9b}$ *19* $\dfrac{3c}{6ac+9bc}$ *20* $\dfrac{3x}{2xy-xz}$ *21* $\dfrac{2pq}{6pr+2pq}$

22 $\dfrac{a+b}{a^2-b^2}$ *23* $\dfrac{x+y}{7x+7y}$ *24* $\dfrac{4a^2-8ab}{a-2b}$ *25* $\dfrac{4a^2-9}{2a+3}$

26 $\dfrac{a^2-4}{a-2}$ *27* $\dfrac{x^2-y^2}{x-y}$ *28* $\dfrac{x+2}{x^2+3x+2}$ *29* $\dfrac{3p-3}{p^2-2p+1}$

$30 \quad \dfrac{ax+5a}{x^2+8x+15}$ $31 \quad \dfrac{4x+20}{3x+15}$ $32 \quad \dfrac{x^2-1}{x^2+2x+1}$ $33 \quad \dfrac{3a^2-5a-2}{a^2-4}.$

$34 \quad \dfrac{x^2-5xy+6y^2}{x^2-4y^2}$ $35 \quad \dfrac{2x^2+x-1}{2x^2+5x-3}$ $36 \quad \dfrac{a^2+2a-35}{a^2+14a+49}$ $37 \quad \dfrac{6c^2-13c+6}{3c^2+10c-8}$

38 Which of the following are true and which are false?

$a \quad \dfrac{px}{qx}=\dfrac{p}{q}$ $b \quad \dfrac{x+2}{x+4}=\dfrac{1}{2}$ $c \quad \dfrac{x+2}{2x+4}=\dfrac{1}{2}$

$d \quad \dfrac{x+2}{x+2}=1$ $e \quad \dfrac{x-2}{2-x}=1$ $f \quad \dfrac{2p+4q+6r}{3p+6q+9r}=\dfrac{2}{3}$

If the numerator or denominator of a fraction is replaced by its negative, the resulting fraction is the negative of the original fraction. Hence, to simplify expressions like

$$\frac{a^2-b^2}{b-a}$$

we can proceed in either of the following ways.

Method 1
The negative of $b-a$ is
$-(b-a)$, i.e. $a-b$, so

$$\frac{a^2-b^2}{b-a}=-\frac{a^2-b^2}{a-b}$$

$$=-\frac{(a-b)(a+b)}{(a-b)}$$

$$=-(a+b)$$

Method 2
The negative of a^2-b^2 is
$-(a^2-b^2)$, i.e. b^2-a^2, so

$$\frac{a^2-b^2}{b-a}=-\frac{b^2-a^2}{b-a}$$

$$=-\frac{(b-a)(b+a)}{(b-a)}$$

$$=-(b+a) \text{ i.e.} -(a+b)$$

Complex fractions

Using the multiplicative identity,

$$\frac{a}{b}=\frac{a}{b}\times1=\frac{a}{b}\times\frac{m}{m}=\frac{am}{bm},\ m\neq0.$$

This basic result enables us to simplify complex fractions in which the numerator or denominator, or both, are fractions.

Example 3

$$\frac{\frac{2}{3}-\frac{1}{4}}{1+\frac{1}{2}} = \frac{12(\frac{2}{3}-\frac{1}{4})}{12(1+\frac{1}{2})}$$

$$= \frac{8-3}{12+6}$$

$$= \frac{5}{18}$$

(In the first line we multiply the numerator and denominator by the LCM of 3, 4 and 2.)

Example 4

$$\frac{\dfrac{1}{y}-\dfrac{1}{x}}{\dfrac{x}{y}-\dfrac{y}{x}} = \frac{xy\left(\dfrac{1}{y}-\dfrac{1}{x}\right)}{xy\left(\dfrac{x}{y}-\dfrac{y}{x}\right)}$$

$$= \frac{x-y}{x^2-y^2}$$

$$= \frac{(x-y)}{(x-y)(x+y)}$$

$$= \frac{1}{x+y}$$

Exercise 8B

Simplify:

1. $\dfrac{a-5}{5-a}$

2. $\dfrac{2p-2q}{q-p}$

3. $-\dfrac{x-y}{y-x}$

4. $\dfrac{ac-ab}{b-c}$

5. $\dfrac{ab-ac}{dc-db}$

6. $-\dfrac{3a-ax}{cx-3c}$

7. $\dfrac{x^2-4}{2-x}$

8. $\dfrac{x^2-4}{2+x}$

9. $\dfrac{x^2-y^2}{y-x}$

10. $\dfrac{1-x^2}{x+1}$

11. $-\dfrac{4x^2-1}{1-2x}$

12. $\dfrac{x^2-9}{6-2x}$

13. $\dfrac{a^2-4}{2a+a^2}$

14. $\dfrac{1-x^2}{2-x-x^2}$

15. $\dfrac{x^4-1}{1-x^2}$

16. $\dfrac{1-\cos^2 A}{1+\cos A}$

17. $\dfrac{4-\sin^2 A}{4+2\sin A}$

18. $\dfrac{3\sin A-3}{1-\sin^2 A}$

19. $\dfrac{1-\cos^2 x}{1-2\cos x+\cos^2 x}$

20. $\dfrac{\cos^2 x-5\cos x+4}{\cos^2 x-16}$

Using a suitable multiplier for numerator and denominator, simplify:

21 $\dfrac{\frac{1}{2}-\frac{1}{3}}{1+\frac{1}{6}}$
  22 $\dfrac{\frac{2}{5}+\frac{1}{2}}{1-\frac{1}{5}}$
  23 $\dfrac{a+\frac{1}{3}}{a-\frac{1}{3}}$

24 $\dfrac{a+\dfrac{1}{b}}{a-\dfrac{1}{b}}$
  25 $\dfrac{1-\dfrac{1}{x}}{1+\dfrac{1}{x}}$
  26 $\dfrac{\dfrac{1}{a}+\dfrac{1}{b}}{a+b}$

27 $\dfrac{1-\dfrac{1}{c^2}}{1-\dfrac{1}{c}}$
  28 $\dfrac{1+\dfrac{x}{y}}{1-\dfrac{x^2}{y^2}}$
  29 $\dfrac{\dfrac{1}{a}-a}{\dfrac{1}{a}+1}$

30 $\dfrac{\dfrac{1}{a^2}-4}{\dfrac{1}{a}-2}$
  31 $\dfrac{\dfrac{1}{x+h}-\dfrac{1}{x}}{h}$
  32 $\dfrac{\dfrac{2s}{c}}{1+\dfrac{s^2}{c^2}}$

33 $\dfrac{\dfrac{1}{x}-\dfrac{1}{x^2}}{\dfrac{1}{x}+\dfrac{1}{x^2}}$
  34 $\dfrac{x-\dfrac{2}{x}+1}{x+\dfrac{1}{x}-2}$
  35 $\dfrac{b-\dfrac{a^2+b^2}{b}}{\dfrac{1}{a}-\dfrac{1}{b}}$

36 Copy and complete the following:

a $2-x = 2\left(\quad-\dfrac{x}{2}\right)$ b $4+x = 4(\quad)$ c $3x-2 = 3(\quad)$

d $4a-2x = 4(\quad)$ e $ax+b = a(\quad)$ f $2x-3x^2 = 2x(\quad)$

---

6 Addition and subtraction of fractions

Two fractions which have the same denominator can be added or subtracted by obtaining the sum or difference of the numerators.

Examples

1. $\dfrac{3}{5}+\dfrac{1}{5}=\dfrac{3+1}{5}=\dfrac{4}{5}$ 2. $\dfrac{3}{5}-\dfrac{1}{5}=\dfrac{3-1}{5}=\dfrac{2}{5}$

3. $\dfrac{a}{c}+\dfrac{b}{c}=\dfrac{a+b}{c}$ 4. $\dfrac{a}{c}-\dfrac{b}{c}=\dfrac{a-b}{c}$

Two fractions which do *not* have the same denominator may be transformed into equivalent fractions which have the same denominator and may then be combined as shown above.

For example, $\frac{2}{5}$ and $\frac{1}{4}$ can be added by finding the LCM of their denominators, 5×4 or 20, and using the fact that

$$\frac{a}{b}=\frac{am}{bm}, \, m\neq0.$$

Then

$$\frac{2}{5}+\frac{1}{4}=\left(\frac{2\times4}{5\times4}\right)+\left(\frac{1\times5}{4\times5}\right)=\frac{8}{20}+\frac{5}{20}=\frac{8+5}{20}=\frac{13}{20}$$

The process may be extended to the sum of three or more fractions.

Example 5. Simplify $\dfrac{a}{b}+\dfrac{c}{d}$

LCM of denominators $=bd$

$$\frac{a}{b}+\frac{c}{d}$$

$$=\frac{ad}{bd}+\frac{bc}{bd}$$

$$=\frac{ad+bc}{bd}$$

Example 6.

Simplify $\dfrac{x+5}{2}-\dfrac{x-2}{3}$

LCM of denominators $=6$

$$\frac{x+5}{2}-\frac{x-2}{3}$$

$$=\frac{3(x+5)}{3\times2}-\frac{2(x-2)}{2\times3}$$

$$=\frac{3(x+5)-2(x-2)}{6}$$

$$=\frac{3x+15-2x+4}{6}$$

$$=\frac{x+19}{6}$$

Exercise 9

Simplify:

1 a $\dfrac{1}{5}+\dfrac{3}{5}$ b $\dfrac{a}{5}+\dfrac{3a}{5}$ c $\dfrac{1}{5a}+\dfrac{3}{5a}$

2 a $\dfrac{1}{4}+\dfrac{1}{2}$ b $\dfrac{a}{4}+\dfrac{a}{2}$ c ' $\dfrac{1}{4a}+\dfrac{1}{2a}$

3 a $\dfrac{4}{5}-\dfrac{2}{5}$ b $\dfrac{4a}{5}-\dfrac{2a}{5}$ c $\dfrac{4}{5a}-\dfrac{2}{5a}$

4 a $\dfrac{1}{2}-\dfrac{1}{3}$ b $\dfrac{a}{2}-\dfrac{a}{3}$ c $\dfrac{1}{2a}-\dfrac{1}{3a}$

5 a $\dfrac{1}{2}+\dfrac{1}{3}+\dfrac{1}{4}$ b $\dfrac{a}{2}+\dfrac{a}{3}+\dfrac{a}{4}$ c $\dfrac{1}{2a}+\dfrac{1}{3a}+\dfrac{1}{4a}$

6 a $\dfrac{1}{2}+\dfrac{1}{3}-\dfrac{3}{4}$ b $\dfrac{a}{2}+\dfrac{a}{3}-\dfrac{3a}{4}$ c $\dfrac{1}{2a}+\dfrac{1}{3a}-\dfrac{3}{4a}$

7   $\dfrac{x}{2}-\dfrac{x}{3}$ 8 $\dfrac{1}{2x}-\dfrac{1}{3x}$ 9 $\dfrac{1}{a}+\dfrac{2}{a}$ 10 $\dfrac{a}{b}+\dfrac{2}{b}$

11   $\dfrac{3}{2a}+\dfrac{1}{2a}$ 12 $\dfrac{5}{2a}-\dfrac{3}{2a}$ 13 $\dfrac{1}{a}+\dfrac{1}{2a}$ 14 $\dfrac{2}{a}+\dfrac{3}{2a}$

15   $\dfrac{2}{9a}+\dfrac{1}{3a}$ 16 $\dfrac{a}{2c}-\dfrac{a}{4c}$ 17 $\dfrac{1}{x}+\dfrac{1}{y}$ 18 $\dfrac{1}{v}-\dfrac{1}{u}$

19   $\dfrac{2}{c}+\dfrac{3}{d}$ 20 $\dfrac{p}{3}-\dfrac{4}{q}$ 21 $\dfrac{a}{x}+\dfrac{1}{y}$ 22 $\dfrac{a}{x}+\dfrac{b}{y}$

23   $\dfrac{a}{b}-\dfrac{c}{d}$ 24 $\dfrac{a}{b}+\dfrac{b}{a}$ 25 $\dfrac{a}{b}-\dfrac{a^2}{b^2}$ 26 $\dfrac{a}{2}+\dfrac{a}{3}-\dfrac{a}{6}$

27   $\dfrac{2}{a}+\dfrac{3}{a}+\dfrac{4}{a}$ 28 $\dfrac{3}{a}+\dfrac{4}{a}-\dfrac{1}{a}$ 29 $\dfrac{x}{2}+\dfrac{2x}{3}+\dfrac{3x}{4}$

30   $1+\dfrac{1}{a}+\dfrac{1}{a^2}$ 31 $\dfrac{1}{c^2}-\dfrac{1}{c}-1$ 32 $\dfrac{1}{a}+\dfrac{1}{b}+\dfrac{1}{c}$

33   $x+\dfrac{1}{x}$ 34 $1-\dfrac{1}{x}$ 35 $a+\dfrac{c}{b}$

36 $\dfrac{x}{2}+\dfrac{x+2}{4}$

37 $\dfrac{x+2}{3}-\dfrac{x+3}{4}$

38 $\dfrac{x-1}{2}-\dfrac{x-2}{5}$

39 $\dfrac{2x-1}{3}-\dfrac{x+2}{4}$

40 $\dfrac{2a-5}{6}-\dfrac{a-7}{3}$

41 $\dfrac{2c-3}{3}+\dfrac{5-3c}{5}$

Example 7. Simplify $\dfrac{2}{x+4}-\dfrac{x-5}{x^2+7x+12}$.

$$\frac{2}{x+4}-\frac{x-5}{x^2+7x+12}=\frac{2}{(x+4)}-\frac{(x-5)}{(x+4)(x+3)}$$

$$=\frac{2(x+3)}{(x+4)(x+3)}-\frac{(x-5)}{(x+4)(x+3)}$$

[The LCM of the denominators is $(x+4)(x+3)$]

$$=\frac{2(x+3)-(x-5)}{(x+4)(x+3)}$$

$$=\frac{2x+6-x+5}{(x+4)(x+3)}$$

$$=\frac{x+11}{(x+4)(x+3)}$$

It is recommended that brackets be inserted as shown in the first line of working.

Exercise 9B

Simplify the following, leaving the denominators in factorised form:

1 $\dfrac{1}{x+1}+\dfrac{1}{x-1}$

2 $\dfrac{3}{a+2}+\dfrac{2}{a+1}$

3 $\dfrac{4}{x+y}+\dfrac{5}{x-y}$

4 $\dfrac{10}{x}-\dfrac{10}{x+1}$

5 $\dfrac{5}{a-1}-\dfrac{5}{a}$

6 $\dfrac{2}{a+b}-\dfrac{3}{a-b}$

7 $\dfrac{x}{x+y}-\dfrac{y}{x-y}$

8 $\dfrac{a}{a-b}-\dfrac{b}{a+b}$

9 $\dfrac{2}{x^2-1}+\dfrac{1}{x+1}$

10 $\dfrac{1}{a^2-1}-\dfrac{1}{a-1}$

11 $\dfrac{3}{x^2+4x+3}+\dfrac{1}{x+3}$

12 $\dfrac{1}{a^2+2a+1}+\dfrac{1}{a+1}$

13 $\dfrac{1}{y-2} - \dfrac{3}{y^2+y-6}$ 14 $\dfrac{1}{a^2-2a+1} + \dfrac{1}{a^2-1}$ 15 $\dfrac{x+4}{x^2-9} - \dfrac{1}{x-3}$

16 $\dfrac{y-1}{y^2-4} + \dfrac{1}{y-2}$ 17 $\dfrac{3}{x^2+x-2} - \dfrac{2}{x^2+3x+2}$ 18 $\dfrac{x+2}{x+3} - \dfrac{x+3}{x+2}$

19 $\dfrac{x-1}{x+1} + \dfrac{x+1}{x-1}$ 20 $\dfrac{1}{x^2} - \dfrac{1}{x^2+x}$ 21 $\dfrac{1}{c^2-c} + \dfrac{1}{c^2}$

22 $\dfrac{2}{6x^2-5x-4} - \dfrac{3}{9x^2-16}$ 23 $\dfrac{1}{\cos A} + \dfrac{1}{\sin A}$ 24 $\dfrac{\sin A}{\cos A} + \dfrac{\cos A}{\sin A}$

25 $\dfrac{1}{1-\sin A} + \dfrac{1}{1+\sin A}$ 26 $\dfrac{\sin x}{1+\cos x} - \dfrac{\cos x}{1-\sin x}$

Summary

1   *Common factors*: $ab + ac = a(b + c)$

Examples
(i) $3x + 9y = 3(x + 3y)$
(ii) $4p^2 - 6p^4 = 2p^2(2 - 3p^2)$
(iii) $x^2yz + xy^2z + xyz^2 = xyz(x + y + z)$

2   *Factors of a difference of squares*: $a^2 - b^2 = (a - b)(a + b)$

Examples
(i) $8 \cdot 5^2 - 1 \cdot 5^2 = (8 \cdot 5 - 1 \cdot 5)(8 \cdot 5 + 1 \cdot 5) = 7 \times 10 = 70$
(ii) $a^2 - 16 = (a - 4)(a + 4)$
(iii) $2x^2 - 18y^2 = 2(x^2 - 9y^2) = 2(x - 3y)(x + 3y)$
(iv) $c^2 - (a + b)^2 = [c - (a + b)][c + (a + b)]$
        $= (c - a - b)(a + b + c)$

3   *Factors of a quadratic expression*: $ax^2 + bx + c$

Examples
(i) $x^2 + 5x + 4 = (x + 4)(x + 1)$
(ii) $a^2 - 4a - 12 = (a - 6)(a + 2)$
(iii) $6y^2 - 7y + 2 = (3y - 2)(2y - 1)$ [OVER

4 Simplification of fractions

(i) $\dfrac{x^2 - x - 6}{2x^2 - 8}$

$= \dfrac{(x-3)(x+2)}{2(x^2-4)}$

$= \dfrac{(x-3)(x+2)}{2(x-2)(x+2)}$

$= \dfrac{x-3}{2(x-2)}$

(ii) $\dfrac{a}{a-b} - \dfrac{b}{a+b}$

$= \dfrac{a(a+b)}{(a-b)(a+b)} - \dfrac{b(a-b)}{(a-b)(a+b)}$

$= \dfrac{a(a+b) - b(a-b)}{(a-b)(a+b)}$

$= \dfrac{a^2 + ab - ab + b^2}{(a-b)(a+b)}$

$= \dfrac{a^2 + b^2}{(a-b)(a+b)}$

(iii) $\dfrac{\dfrac{a}{b} - \dfrac{b}{a}}{\dfrac{1}{b} + \dfrac{1}{a}}$

$= \dfrac{ab\left(\dfrac{a}{b} - \dfrac{b}{a}\right)}{ab\left(\dfrac{1}{b} + \dfrac{1}{a}\right)}$ *multiplying numerator and denominator by ab*

$= \dfrac{a^2 - b^2}{a+b}$

$= \dfrac{(a-b)(a+b)}{(a+b)}$

$= a - b$

(iv) If the numerator, or denominator, of a fraction is replaced by its negative, then the original fraction is replaced by its negative.

Replacing $b - a$ by its negative

$- \dfrac{a-b}{b-a}$

$= + \dfrac{a-b}{a-b}$

$= 1$

The Language of Variation

1 Direct variation, by formula

Suppose we wish to find out how the stretch in the length of a piece of elastic depends on the pull applied to it. One way to tackle this problem is to set up an experiment. The piece of elastic is fastened to a support by a drawing pin, and the stretch is measured for different pulls. The pull is provided by the weight of a known mass in a scale pan which is attached to the free end of the piece of elastic (see Figure 1).

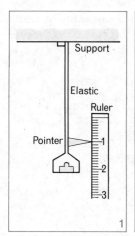

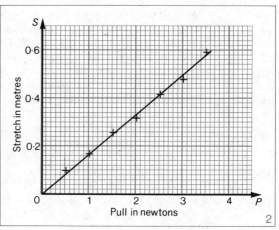

Figure 2 shows the graph of the experimental results. The points seem to lie on a straight line through the origin, allowing for slight deviations due to experimental error. Part of this line has been drawn.

27

Here are some pairs of corresponding readings, *taken from the line*:

Pull in newtons (P)	0·6	1·2	1·8	2·4	3·0	3·6
Stretch in metres (S)	0·1	0·2	0·3	0·4	0·5	0·6

Notice that for each ordered pair (*P*, *S*), the ratio

$$\frac{S}{P} = \frac{1}{6}$$

so we can write $S = \frac{1}{6}P$.

Alternatively, since the graph is a straight line through the origin, its equation is $S = mP$, where *m* is the gradient of the line. From the graph, the gradient is

$$\frac{0·6}{3·6} = \frac{1}{6}$$

giving the same result as before. Hence, for this piece of elastic, we can say that the *formula*, or *law*, relating the stretch *S* metres to the pull *P* newtons is $S = \frac{1}{6}P$.

When you met this kind of problem in Arithmetic, Book 3, you recognised it to be *direct proportion* and solved problems by using either the *rate* method or the *ratio* method. In this context, if the pull is doubled, then the corresponding stretch is doubled; if the pull is halved, the corresponding stretch is halved.

In this chapter, the ideas of proportion will be extended, and for this reason it will be convenient to extend our mathematical language also. Thus, the direct proportion illustration above can be described in the language of variation as follows:

(i)   *S varies directly as P*, or simply *S varies as P*.
(ii) $S \propto P$ (read as in (i)).
(iii) $S = kP$, where *k* is a constant, called the *constant of variation*, which is determined from the data.
(iv) The stretch varies as the pull.

These four statements mean exactly the same thing.

Example. The volume V cm^3 of a gas varies directly as its absolute temperature T K under certain conditions. If $V = 1500$ when $T = 300$, find V when $T = 500$.

$$V \propto T$$

i.e. $\quad V = kT$

$\Rightarrow 1500 = 300k \quad$ (The symbol $\Rightarrow$ means 'implies'.)

$\Rightarrow \quad k = 5$

$\Rightarrow \quad V = 5T$

When $\quad T = 500, \quad V = 5 \times 500$

$$= 2500$$

Note. $V = 5T$ is the *formula* connecting V and T.

Exercise 1

1   In each of these sets of ordered pairs (x, y) fill in the blanks in such a way that the ratio $y : x$ is constant:

a $\{(3, 18), \quad (7, 42), \quad (2\frac{1}{2}, 15), \quad (8, ...), \quad (..., 30)\}$

b $\{(3, 6), \quad (7, 14), \quad (9, ...), \quad (..., 2\cdot4)\}$

c $\{(1, 0\cdot2), \quad (10, 2), \quad (30, ...), \quad (45, ...), \quad (..., 2\cdot4)\}$.

2   Complete the following table on the assumption that $y \propto x$:

x	2	5	8	...	...
y	...	20	...	40	7

State the constant of variation, and give the formula relating y and x in the form $y = kx$.

3   Use the formula $y = \frac{3}{2}x$ to complete these ordered pairs:

$(6, ...), \quad (7, ...), \quad (-4, ...), \quad (9, ...), \quad (..., 21), \quad (..., -12)$.

4   Write each of the following in the forms $y \propto x$ and $y = kx$, using the symbols given:

a   The perimeter P units of an equilateral triangle varies directly as the length of a side x units.

b   The volume V cm^3 of a perfect gas varies directly as its absolute temperature T K.

c   The circumference c units of a circle varies as the radius r units.

d   The distance s km travelled by a car at a steady speed varies as the time t hours.

e The cost £*P* of transporting goods varies as the distance *d* km.

5 If $y = kx$, and $y = 15$ when $x = 6$, find the constant k. Find y when $x = 10$.

6 y varies directly as x, and so $y = kx$. If $y = 6$ when $x = 3$, find the constant of variation k. Hence calculate y when $x = 11$.

7 If $y \propto x$ so that $y = kx$, and $y = 4$ when $x = 8$, find y when $x = 9$.

8 If y varies directly as x, and $y = 12$ when $x = 3$, find an equation relating y and x. Use this equation to find:
 a y when $x = 9$ *b* x when $y = 52$.

9 If y varies directly as x, and $y = 24$ when $x = 16$, find the constant of variation.
  Hence calculate: *a* y when $x = 24$ *b* x when $y = 33$.
  If x is trebled, what is the effect on y?

10 If x varies as y, and $x = 3.6$ when $y = 1.5$, find x when $y = 5$.

11 The tension T newtons in a spring varies as the extension e centimetres (Hooke's law). A tension of 24 newtons stretches a certain spring by 4·5 cm. Find the formula connecting T and e, and use it to calculate T when $e = 2.4$.

12 The force P newtons required to pull a load of weight W newtons along a horizontal surface is such that P varies as W. Given that $P = 90$ when $W = 375$, find an equation connecting P and W. Calculate: *a* P when $W = 120$ *b* W when $P = 72$.

13 In silver plating, the number of grammes m of silver deposited on a metallic surface varies as the number of hours t, the electric current being kept steady. Given that $m = 1.4$ when $t = 0.5$, find an equation giving m in terms of t. Hence find the number of grammes that will be deposited in 2·5 hours.

14 In an electrical experiment, the results showing the relation between the voltage V volts and the current I amperes were as follows:

I	0·1	0·2	0·3	0·4	0·5	0·6
V	2	4	6	8	10	12

a Show these results on a coordinate graph, and give a reason why the probable formula relating I and V is $V = kI$.

b Using your graph, or the table, find the constant k and hence state the formula.

c Use the formula to calculate:

(*1*) V when $I = 0.45$ (*2*) I when $V = 6.4$.

---

2 *Direct variation, using a slide rule*

(*i*) *The rate method*

If (x_1, y_1), (x_2, y_2), (x_3, y_3), ... are ordered pairs which satisfy the equation $y = kx$, we can equally well say that

$$\frac{y_1}{x_1} = \frac{y_2}{x_2} = \frac{y_3}{x_3} = ... = k$$

This description of direct variation by the *rate method*, which we have been using in Section 1, is well suited to the slide rule. By setting y_1 on the D scale against x_1 on the C scale, the other pairs of corresponding numbers can be read off at once as indicated in Figure 3.

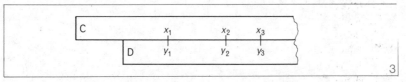

3

If CF and DF (folded) scales are not available, there will be a chance of 'running off the scale', which necessitates the resetting of the slide. To avoid this, B and A scales can be used instead of C and D; the accuracy is then halved.

Example 1. In a French test, the possible mark is 65. Convert a mark of 39 out of 65 to a percentage mark.

Old mark		New mark
65		100
39		x

Set 65 on the C scale against 100 on the D scale. Opposite 39 on the C scale is 60 on the D scale and so the percentage mark is 60.

(ii) The ratio method

There is another way of looking at direct variation. If (x_1, y_1), (x_2, y_2) are ordered pairs which satisfy the equation $y = kx$, then

$$y_1 = kx_1$$
$$and \quad y_2 = kx_2$$

Dividing, $\quad \dfrac{y_2}{y_1} = \dfrac{kx_2}{kx_1} \quad$ i.e. $\quad \dfrac{y_2}{y_1} = \dfrac{x_2}{x_1}$

which is the *ratio method* for direct proportion.

Example 2. The cost of petrol for a car journey varies directly as the number of kilometres. If petrol for a journey of 380 km costs £2·60, estimate the cost of petrol for 640 km.

Let £c be the cost of petrol for a journey of x km. Then $c = kx$. The data may be tabulated as follows:

Distance x km	Cost £c
380	2·60
640	c

Using the ratio method, set 380 on the C scale against 640 on the D scale. Opposite 2·60 on the C scale is 4·38 on the D scale so that the estimated cost is £4·38.

Check this result by the rate method.

Exercise 2 (using a slide rule)

1   Assuming that y varies as x, complete this table:

x	16	17	18	19	20	21	22	...	...
y	40	...	...	...	...	...	...	62·5	74

2   Assuming that I varies as e, complete the following table:

e	4·5	5·0	5·5	6·0	6·5	7·0	...	...
I	24	...	...	...	...	...	41·6	46·4

3   Complete the following table:

Miles	1	5	10	15	20	25	30	100
Kilometres	1·61	...	...	...	...	...	...	...

4   Convert the following marks, out of a possible 80, to percentage marks.

Mark out of 80	27	43	44	50	52	54	57	63	76
Percentage mark	...	...	...	...	...	...	...	...	...

5   The tension T newtons in a spring varies directly as the extension x millimetres. Use this information to complete the following table:

x	0·2	0·4	0·5	...
T	50	100	...	175

6 *a*   Calculate x, given that $\dfrac{x}{15 \cdot 4} = \dfrac{4 \cdot 6}{5 \cdot 8}$

   b   Find y such that $\dfrac{13 \cdot 5}{y} = \dfrac{27}{19}$

7     $y \propto x$. Given that $y = 3 \cdot 8$ when $x = 2 \cdot 2$, calculate:

 a y when $x = 4 \cdot 5$ *b* x when $y = 10 \cdot 7$.

8   Within slide-rule accuracy, write down the set into which $\{3 \cdot 4, 8 \cdot 5, 11 \cdot 7\}$ is mapped under the mapping $x \to \frac{4}{13} x$.

9 ·   The petrol gauge markings of a car show when the tank holds 10, 20, 30, 40 litres. If 1 gallon is equivalent to 4·55 litres, show in a table the number of gallons that correspond to these markings.

10   In any triangle, the lengths of the sides are directly proportional to the sines of the opposite angles. Given that $a = 4 \cdot 5$, $\sin A = 0 \cdot 60$ and $\sin B = 0 \cdot 84$, calculate b.

     The constant of proportionality is equal to $2R$, where R is the radius of the circle through the three vertices. Calculate R.

---

3 *Other forms of direct variation*

For the 'pull and stretch' investigation of Section 1, the graph is a much better representation of the relation than the table, which shows only some of the possible ordered pairs.

If a relation is of the type 'y varies as x', its graph is a straight line through the origin, and *conversely*, if the graph of y against x is a straight line through the origin, y varies as x. Furthermore, the graph can be used to find the constant of variation since this gives the gradient of the line.

Example 1. A ball-bearing, starting from rest, is allowed to roll down a straight channel which is inclined at an angle to the horizontal. The distance, d metres, travelled in t seconds is given below:

t	0	0·5	1·0	1·5	2·0	2·5	3·0
d	0	0·5	2·0	4·5	8·0	12·5	18·0

It is evident that the ratio $d:t$ is not constant, so that d does not vary as t; this is confirmed by the graph in Figure 4.

Notice that the graph resembles the graph of the mapping $x \rightarrow x^2$ (drawn in Book 5, Chapter 3). To test this, we construct a table of ordered pairs (t^2, d) and draw a graph of d against t^2.

t^2	0	0·25	1·00	2·25	4·00	6·25	9·00
d	0	0·5	2·0	4·5	8·0	12·5	18·0

Since the graph in Figure 5 is a straight line, we deduce that d varies as t^2, and so $d = kt^2$.

From the table (or graph), $k = 2$, and hence the formula is $d = 2t^2$.

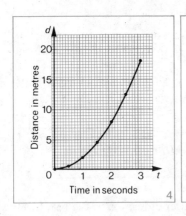

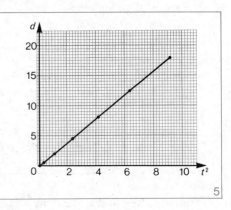

In general, y is said to vary directly as the nth power of x, written $y \propto x^n$, if $\dfrac{y}{x^n}$ is a constant, i.e. if $y = kx^n$, where k is a constant.

Example 2. The distance of the visible horizon varies as the square root of the height of the observation point above sea-level. At a height of 4 metres it is known that the distance is 5·2 kilometres. Find the distance at a height of 9 metres.

Let the distance be d km at a height of h m. Then the *law of variation* is $d \propto \sqrt{h}$, i.e. $d = k\sqrt{h}$.

(i) *Method of Section 1*

$$d = k\sqrt{h}$$
$$\Rightarrow 5{\cdot}2 = k\sqrt{4}$$
$$\Rightarrow k = 2{\cdot}6$$
$$\Rightarrow d = 2{\cdot}6\sqrt{h}$$

When $h = 9$, $d = 2{\cdot}6\sqrt{9}$
$$= 7{\cdot}8$$

The distance is 7·8 kilometres.

(ii) *Ratio method of Section 2*

$$d_1 = k\sqrt{h_1} \text{ and } d_2 = k\sqrt{h_2}$$
$$\Rightarrow \frac{d_2}{d_1} = \frac{\sqrt{h_2}}{\sqrt{h_1}}$$

Data

d	h
5·2	4
d_2	9

Hence $\dfrac{d_2}{5{\cdot}2} = \dfrac{\sqrt{9}}{\sqrt{4}} = \dfrac{3}{2}$

$$\Rightarrow 2d_2 = 3 \times 5{\cdot}2$$
$$\Rightarrow d_2 = 7{\cdot}8$$

Figures 6 and 7 show the graphs of d against h and d against $\sqrt{h}$.

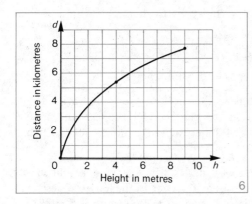

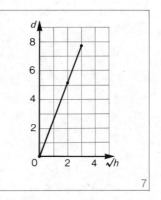

Exercise 3

1   Express each of the following as an algebraic equation, denoting the constant of variation in each case by k:

   a $p \propto q^2$ *b* $r \propto \sqrt{A}$ *c* $W \propto d^3$ *d* $D \propto \tan x$

2   Express each of the statements in question *1* in words, using the language of variation.

3   y varies directly as x^2, and $y = 12$ when $x = 2$. What is the constant of variation? Find:

   a y when $x = 5$ *b* x when $y = 0.48$, $x > 0$.

4   y varies as the square root of x, and $y = 1$ when $x = 4$. Find:

   a y when $x = 36$ *b* x when $y = 5$.

5   $y \propto x^3$, and $y = 4$ when $x = 1$. Find:

   a y when $x = 5$ *b* x when $y = 108$.

6   y varies as the square of x, and $y = 1.6$ when $x = 2$. Find y when $x = 10$.

7   If y varies as $(x + 1)$, and $y = 3$ when $x = 5$, show that $2y - x = 1$.

8   If $x \propto y^2$, and $x = a^2$ when $y = 2a$, show that $y^2 = 4x$.

9   The resistance R newtons to the motion of a train varies as the square root of its speed V metres per second. If the resistance is 3600 newtons when the speed is 25 metres per second, calculate R when $V = 100$.

10   For a light metal beam of length L m, carrying a load at its midpoint, the sag S cm varies as the cube of the length. Given that $S = 2$ when $L = 4$, calculate the sag when the length is 8 metres.

11   For question *10*, calculate the length of a similar beam, carrying the same load, when the sag is 0.25 cm.

12   The safe speed v metres per second at which a train can round a curve of radius r metres varies as $\sqrt{r}$. If the safe speed for a radius of 100 m is 20 metres per second, calculate the safe speed for a radius of 95 m.

13 The distance s metres through which a heavy body falls from rest varies as the square of the time, t seconds, taken. When $t = 2$, $s = 20$ approximately. How far will the body fall from rest in 6 seconds?

14 In question 13, estimate how long the body would take to fall 45 metres from rest.

15 The time of swing T seconds of a pendulum varies as the square root of the length L centimetres of the pendulum. When the length is 81 cm the time of swing is 1·8 seconds. Calculate the time of swing for a pendulum 50 cm long.

16 The square of the time (T years) taken by a planet to revolve round the sun varies as the cube of the distance (D km) of the planet from the sun. The earth takes 1 year, at a distance of about 150 million km. Calculate the time of revolution of Mars, which is about 225 million km from the sun. (Do not simplify the value of k.)

17 The areas of similar triangles are proportional to the squares on corresponding sides. If x_1 cm and x_2 cm are the lengths of the corresponding sides of two similar triangles whose areas are A_1 cm^2 and A_2 cm^2 respectively, show that $\dfrac{A_1}{A_2} = \dfrac{x_1^2}{x_2^2}$.

a If $x_1 = 3$ and $x_2 = 2$, state the ratio of the areas.
b If $A_2 = 22\cdot5$, $x_1 = 4$ and $x_2 = 6$, find A_1.

18 $y \propto x^2$. What is the effect:
a on y, if x is doubled
b on y, if x is reduced to one third of its value?

19 Repeat question 18 in the case where $y \propto x^3$.

20 Given that y varies as x^n, where n is a positive integer, find a simple law of variation in the form $y = kx^n$ for the following, and complete the table:

x	1	2	3	...
y	2	8	...	50

21 The rate, V cm^3 per second, at which water flows from a valve at the base of a tank when the depth of water is h cm is given in the table:

h	0	4	9	16	25	36
V	0	28·4	42·6	56·8	71·0	85·2

a Show the data graphically, using a scale of 2 cm to 5 units for h and 2 cm to 10 units for V. Deduce that V does not vary as h.

b Draw another graph on the same diagram, showing the relation between V and $\sqrt{h}$. Hence show that $V = k\sqrt{h}$, and use your graph to find k.

c Calculate V when $h = 20$.

4 Inverse variation

Example 1. The table shows some ordered pairs from the one-to-one correspondence between times and average speeds for a distance of 120 m.

Time, t seconds	1	2	3	4	5	6	8	10	12
Speed, v m/s	120	60	40	30	24	20	15	12	10

It is clear that v does not vary as t. The graph in Figure 8 verifies this.

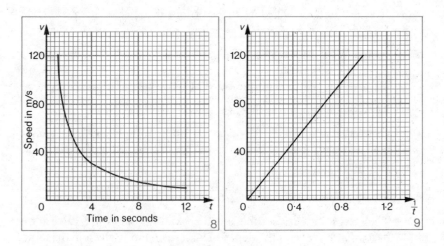

The graph resembles that of the mapping $x \to 1/x$ (a rectangular hyperbola). This suggests that we draw the graph of v against $1/t$.

Figure 9 shows the graph and the table shows ordered pairs $(1/t, v)$:

$1/t$	1	0·50	0·33	0·25	0·20	0·17	0·13	0·10	0·08
v	120	60	40	30	24	20	15	12	10

This graph is a straight line through the origin; we deduce that v varies as $1/t$ and so $v = k/t$, i.e. $vt = k$. In the language of variation, we say that:

$$v \text{ varies inversely as } t, \quad v \propto \frac{1}{t}, \quad v = \frac{k}{t}, \text{ or } vt = k.$$

You will recognise this as *inverse proportion*. Methods for solving problems correspond to the methods used for solving problems in inverse proportion in Arithmetic, Book 3.

Example 2. In Example 1, v varies inversely as t, and $v = 30$ when $t = 4$. Calculate v when $t = 9$.

Method 1

$$v = \frac{k}{t}$$

$$\Rightarrow 30 = \frac{k}{4}$$

$$\Rightarrow k = 120$$

$$\Rightarrow v = \frac{120}{t}$$

When $t = 9$, $v = \dfrac{120}{9}$

$$= 13\tfrac{1}{3}$$

Method 2

If (t_1, v_1) and (t_2, v_2) are ordered pairs which satisfy the equation $vt = k$, then $v_1 t_1 = k$ and $v_2 t_2 = k$

$$\Rightarrow v_1 t_1 = v_2 t_2$$

$$\Rightarrow \frac{v_2}{v_1} = \frac{t_1}{t_2}$$

Hence $\dfrac{v_2}{30} = \dfrac{4}{9}$

Data	
v	t
30	4
v_2	9

$$\Rightarrow 9v_2 = 120$$

$$\Rightarrow v_2 = \tfrac{120}{9}$$

$$= 13\tfrac{1}{3}$$

Note. In method 2, the ratio $v_2/30 = 4/9$ is well suited for calculation using a slide rule. Set 9 on the C scale opposite 4 on the D scale, and read 13·3 on the D scale opposite 30 on the C scale.

Exercise 4

1   If $y = 4/x$ give the relation between the variables in words. Find:

   a y when $x = 2$ b x when $y = 0.1$.

2   y varies inversely as x, and $y = 3$ when $x = 4$. Find an equation connecting y and x. Hence find y when $x = 6$.

3   Given that $y \propto 1/x$, and $y = 6$ when $x = 14$, obtain an equation in x and y. Find y when $x = 28$.

4     $p \propto 1/q$, and $p = 6$ when $q = 5$. Find p when $q = 12$.

5     $zx = k$, where k is a constant, and $z = 18$ when $x = 2\cdot5$. Find z when $x = 3\cdot6$.

6   y varies inversely as x, and $y = 8$ when $x = 3$. Find:

  a   an equation connecting y and x
  b y when $x = 60$ c x when $y = 15$.

7   The pressure P pascals (newtons per square metre) of a given mass of gas at constant temperature varies inversely as the volume V m^3. When $P = 600$, $V = 2$. Find:

  a P when $V = 3$ b V when $P = 800$.

8   At a curve of radius R metres on a railway, the outer rail has to be raised above the level of the inner rail by a height H cm. It is known that H varies inversely as R.

  a   For $R = 500$, $H = 12\cdot8$. Find a formula connecting H and R.
  b   Calculate H when $R = 480$.

9   The frequency f kHz of radio waves is inversely proportional to their wavelength λ metres. $f = 375$ when $\lambda = 800$.

  a   Find a formula connecting f and λ.
  b   Calculate the wavelength of Radio 2 whose frequency is 200 kHz.
  c   Find, to the nearest unit, the frequency of Radio 3 whose wavelength is 464 m.

10   The following table shows the variation of y with respect to x:

x	1	2	3	4	6	10
y	6	3	2	1·5	1	0·6

  a   Make up a new table showing how y varies with $1/x$, giving results correct to two decimal places.
  b   Using a scale of 1 cm to 0·1 unit for $1/x$ and 1 cm to represent 0·5 unit for y, draw the graph of y against $1/x$.

c From your graph, give a reason why y varies inversely as x. Find the value of the constant of variation.

d Find y when $x = 5$.

---

5 *Other forms of inverse variation*

In the same way that the language of variation extends the idea of direct variation, $y \propto x$, to other laws of variation such as $y \propto x^2$, $y \propto \sqrt{x}$, $y \propto x^3$, it also extends the idea of inverse variation, $y \propto 1/x$, to $y \propto 1/\sqrt{x}$, $y \propto 1/x^2$, etc. The last mentioned is a particularly important one, called the *inverse square law*, which is of importance in science. Instances are given in questions *8* and *11* of Exercise 5 and also in the worked example below.

In general, y is said to vary *inversely* as the nth power of x, written $y \propto 1/x^n$, if $yx^n = k$, or $y = k/x^n$, where k is a constant.

Example. Two electrical charges attract one another with a force, P units, which varies inversely as the square of the distance, x units, between them. $P = 5.4$ when $x = 6$; find P when $x = 9$.

Method 1

$$P = \frac{k}{x^2}$$

$$\Rightarrow 5.4 = \frac{k}{6^2}$$

$$\Rightarrow k = 194.4$$

$$\Rightarrow P = \frac{194.4}{x^2}$$

When $x = 9$, $P = \dfrac{194.4}{81}$

$$= 2.4$$

Method 2

$$P_1 = \frac{k}{x_1^2} \quad \text{and} \quad P_2 = \frac{k}{x_2^2}$$

$$\Rightarrow \frac{P_2}{P_1} = \frac{x_1^2}{x_2^2}$$

Hence $\dfrac{P_2}{5.4} = \dfrac{6^2}{9^2} = \dfrac{4}{9}$

$$\Rightarrow 9P_2 = 4 \times 5.4$$

$$\Rightarrow P_2 = 2.4$$

Data

P	x
5.4	6
P_2	9

Exercise 5

1   Express each of the following as an algebraic equation:

    a $p \propto \dfrac{1}{v}$ *b* $y \propto \dfrac{1}{\sqrt{x}}$ *c* $y \propto \dfrac{1}{x^3}$

2   Express each of the statements in question *1* in words, using the language of variation, and in particular the word 'inversely'.

3   *y* varies inversely as x^2, and $y = 9$ when $x = 2$. Find:

    a *y* when $x = 10$ *b* *x* when $y = 4$.

4    $y \propto 1/\sqrt{x}$, and $y = 5$ when $x = 9$. Find *y* when $x = 25$.

5   Given that *y* varies inversely as the square of *x*, and $y = 48$ when $x = 3$, calculate *y* when $x = 4$.

6   *y* is inversely proportional to $\sqrt{x}$. If $y = 5$ when $x = 16$, find:

    a *y* when $x = 100$ *b* *x* when $y = 60$.

7    $y \propto 1/x^3$, and $y = 10$ when $x = 2$. Find *y* when $x = 10$.

8   The intensity of illumination, *I* lumens per square metre, at a point on a screen varies inversely as the square of the distance, *d* metres, of the light source from the point. When $d = 5$, $I = 1$.

    a Express *I* in terms of *d*.   *b* Find *I* when $d = 2 \cdot 5$.
    c Calculate *d* when $I = 2 \cdot 5$.

9   For a given volume, the height, *h* cm, of a cone varies inversely as the square of the radius, *r* cm, of the base. When $r = 5$, $h = 8$. Find the height of a cone of radius 4 cm.

10   The time of vibration, *T* seconds, of a loaded beam is inversely proportional to the square root of the deflection, *x* mm, produced by the loading. When $x = 1 \cdot 00$, $T = 0 \cdot 20$. Find:

    a *T* when $x = 0 \cdot 64$ *b* *x* when $T = 0 \cdot 50$.

11   The weight, *W* newtons of a body varies inversely as the square of its distance, *x* km, from the centre of the earth. If *R* km is the radius of the earth and W_0 is the weight of the body at the surface of the earth, show that:

$$W = \frac{R^2 W_0}{x^2}.$$

Hence estimate the weight, in a space capsule 5600 km *above the surface of the earth*, of a man weighing 900 newtons on earth, given that R is approximately 6400.

12 Given that $y \propto 1/x^2$, what is the effect on y:
  a if x is doubled   b if x is halved?

6 Variation involving several variables

The volume V of a cylinder with radius r and height h is given by the formula $V = \pi r^2 h$, and depends on both r and h, each of which can be changed independently of the other.

(i)   If r is multiplied by 3, V is multiplied by 3^2, i.e. 9, and so if h is constant, $V \propto r^2$ (see Figure 10(i)).

(ii) If h is multiplied by 2, V is multiplied by 2 and so if r is constant, $V \propto h$ (see Figure 10(ii)).

(iii) If r is multiplied by 3, and h is multiplied by 2, V is multiplied by $3^2 \times 2$, i.e. 18, and we say that V varies *jointly* as r^2 and h, i.e. $V \propto r^2 h$ and so $V = kr^2 h$ (see Figure 10(iii)). This is known as *joint variation*.

Note. The word 'jointly' is frequently omitted from the statement.

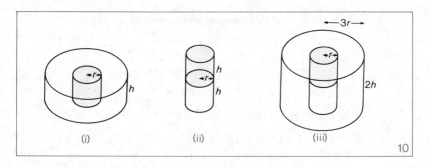

In general, y is said to vary *jointly* as x and z, written $y \propto xz$, if $y = kxz$, where k is the constant of variation.

The ideas of joint variation can be extended as in the following example.

Example. y varies as x^2 and inversely as z. $y = 12$ when $x = 6$ and $z = 9$. Find y when $x = 9$ and $z = 12$.

Method 1

$$y = \frac{kx^2}{z}$$

$$\Rightarrow 12 = \frac{k \times 6^2}{9}$$

$$\Rightarrow k = \frac{9 \times 12}{36} = 3$$

$$\Rightarrow y = \frac{3x^2}{z}$$

When $x = 9$ and $z = 12$,

$$y = \frac{3 \times 9^2}{12}$$

$$= 20 \cdot 25$$

Method 2

$$y_1 = \frac{kx_1^2}{z_1} \quad \text{and} \quad y_2 = \frac{kx_2^2}{z_2}$$

$$\Rightarrow \quad \frac{y_2}{y_1} = \frac{x_2^2}{x_1^2} \cdot \frac{z_1}{z_2}$$

Hence $\quad \dfrac{y_2}{12} = \dfrac{9^2}{6^2} \times \dfrac{9}{12} = \dfrac{27}{16}$

$$\Rightarrow 16y_2 = 12 \times 27$$

$$\Rightarrow \quad y_2 = \frac{12 \times 27}{16}$$

$$= 20 \cdot 25$$

Data

y	x	z
12	6	9
y_2	9	12

Note. So far we have been using symbols for measures of quantity, i.e. for numbers. But often in the language of variation the symbols are used also for the quantities themselves. This practice is followed in the next Exercise; remember that *the constant of variation*, if used, *depends on the units*.

Exercise 6

1   Express each of the following algebraically as an equation:

a A varies jointly as r and h.
b p varies as t and inversely as v.
c Q varies jointly as x and the square of y.
d F varies as m_1 and m_2 and inversely as the square of d.

2   x varies jointly as y and z. $x = 24$ when $y = 3$ and $z = 2$.

a   Find an equation connecting x, y and z.
b   Find x when $y = 5$ and $z = 4$.

3   y varies as x and z, and $y = 12$ when $x = 2$ and $z = 3$.

a   Obtain an equation connecting x, y and z.
b   Find y when $x = 4$ and $z = 5$.
c   Find x when $y = 18$ and $z = 9$.

4    $p \propto qr$, and $p = 18$ when $q = 4$ and $r = 9$. Find p when $q = 6$ and $r = 2\cdot5$.

5   F varies as y and inversely as z, and $F = 8$ when $y = 4$ and $z = 3$.

    a Find a formula for F.
    b Calculate F when $y = 5$ and $z = 2$.

6   Q varies as x^2 and inversely as y. If $Q = 1$ when $x = 3$ and $y = 18$, find Q when $x = 10$ and $y = 32$.

7   T varies directly as x and inversely as the square root of y. $T = 12$ when $x = 6$ and $y = 4$. Find T when $x = 15$ and $y = 9$.

8   y varies directly as x and z and inversely as the square of t. $y = 100$ when $x = 25$, $z = 2$ and $t = 1$. Find

    a y when $x = 12$, $z = 5$ and $t = 4$
    b x when $y = 36$, $z = 3$ and $t = 2$.

9   The weight W of a bar varies jointly as its length L and the square of its diameter d. Given that $W = 640$ when $L = 120$ and $d = 8$, find a formula for W in terms of L and d. Calculate:

    a W when $L = 160$ and $d = 9$ *b* L when $W = 240$ and $d = 4$
    c d when $W = 150$ and $L = 50$.

10   The conductance G of a solution varies directly as the area A of the cross-section and inversely as the length L. $G = 1\cdot2$ when $A = 1\cdot5$ and $L = 3$. Find G when $A = 2\cdot0$ and $L = 3\cdot6$.

    Example. V varies directly as r^2 and h so that $V \propto r^2 h$.
    a Find the effect on V if r is trebled and h is halved.
    b What is the percentage change in V if r is increased by 20% and h is decreased by 25%?

    a If r is multiplied by 3 and h is multiplied by $\frac{1}{2}$, then V is multiplied by $3^2 \times \frac{1}{2}$, i.e. $4\cdot5$.
    b If r is increased by 20% and h is decreased by 25%, then the percentage change in V is

$$\left(\frac{120}{100}\right)^2 \times \frac{75}{100}, \quad \text{i.e.} \quad \frac{36}{25} \times \frac{3}{4} \quad \text{or} \quad \frac{108}{100}.$$

Hence V is increased by 8%.

Exercise 6B

1   y varies jointly as x and z. $y = 42$ when $x = 5$ and $z = 2\cdot4$.

   a Find a formula for y.
   b Calculate y for $x = 2$ and $z = 9$.

2   x varies as the square of y and inversely as z. $x = 20$ when $y = 4$ and $z = 12$. Find x when $y = 1\cdot2$ and $z = 3$.

3   y varies directly as x^2 and inversely as $\sqrt{z}$. Find y in terms of x and z given that $y = 8$ when $x = 2$ and $z = 25$. Calculate y when $x = 3$ and $z = 1$.

4   Q varies as p and q and inversely as t. Given that $Q = 1$ when $p = 3$, $q = 4$ and $t = 6$, calculate Q for $p = 4$, $q = 9$ and $t = 12$.

5   The volume V of a given mass of gas varies as the absolute temperature T and inversely as the pressure p. At temperature $360\,\text{K}$ and pressure $736\,\text{mmHg}$, the volume is $450\,\text{cm}^3$.

   a Find a formula for V.
   b Find V when the temperature is $312\,\text{K}$ and the pressure is $960\,\text{mmHg}$.

6   The pressure p on a disc immersed in a liquid varies as the depth d and the square of the radius r of the disc. The pressure is $3000\,\text{Pa}$ (newtons per square metre) when the depth is 3 m and the radius is 2 m.

   a Find a formula for p.
   b Calculate p when the depth is 4 m and the radius is $1\cdot5$ m.
   c Calculate d when the pressure is $6000\,\text{Pa}$ and the radius is 4 m.

7   The force F to stop a train varies as the square of the speed V and inversely as the stopping distance S. A force of $100\,\text{kN}$ will stop a certain train travelling at 15 m/s in 270 m.

   a What force would stop the same train in 400 m when travelling at 24 m/s?
   b If a force of $180\,\text{kN}$ brings the train to a halt in 600 m, at what speed was it travelling initially?

8   The electrical resistance R of a wire varies directly as the length L and inversely as the square of its diameter d. $R = 1\cdot8$ when $L = 150$ and $d = 0\cdot05$. Calculate R when $L = 90$ and $d = 0\cdot03$.

9   The potential V at a point distant r from a point charge q varies as q and inversely as r. The potential $V = 40$ when $q = 10^{-8}$ and $r = 0.5$. Find V when $q = 4 \times 10^{-8}$ and $r = 0.4$.

10   P varies jointly as a^2 and b, and so $P \propto a^2 b$.

a   If a is doubled and b is halved, what is the effect on P?

b   If a is increased by 25% and b is decreased by 4%, what is the percentage change in P?

11   Q varies directly as the square of x and inversely as t. If x is increased by 20% and t is decreased by 10%, find the percentage change in Q.

12   The safe load W of a beam supported at each end varies as the breadth b and the square of its depth d, and inversely as the distance x between the supports.

a $W = 8400$ when $b = 7.5$, $d = 15$ and $x = 5$. Find W when $b = 6$, $d = 12$ and $x = 4$.

b   If b is increased by 50%, d is decreased by 50% and x is increased by 25%, what is the percentage change in W?

7 The linear law from experimental data

The formulae, or laws, of direct and inverse variation contain only one constant which can be found by the methods of Sections 1–6, without constructing a graph. The experimenter is often confronted by a different situation in which the variables may be connected by an equation containing more than one constant. In this case, it is best to draw a graph of the relation using ordered pairs from the experimental data. The formula, or law, may then be found by the method shown in the following example.

Example. The volume of water, V litres, in a tank after t seconds when water is flowing from an outlet at the base is given below:

t	5	10	15	20	25
V	128	100	76	52	24

It is suspected that a law of the form $V = at + b$ exists. Find if this is the case, and if so calculate a and b, and give the formula.

The points are shown in Figure 11, and a '*best-fitting*' straight line has been drawn.

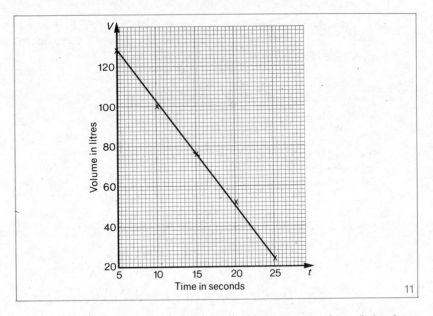

11

We know that every straight line has an equation of the form '$y = mx + c$' for the usual Cartesian axes and coordinates. In the above example the variables are t and V, and since the points lie approximately on a straight line, a law of the form $V = at + b$ exists, where a and b are constants.

Since the graph of the relation is a straight line, the law is sometimes called a *straight-line law* or a *linear law*. In the language of variation, we say that y varies *partly* as x and is *partly* constant.

We now find the linear law for the worked example. Choosing two convenient well-separated points *on the line*, we have:

$$t = 5, \ V = 128 \ \Rightarrow \ 5a + b = 128$$

$$t = 25, \ V = 24 \ \Rightarrow \ 25a + b = 24$$

Subtract. $\quad -20a = 104$

$$\Leftrightarrow \quad a = -5 \cdot 2$$

Substituting $a = -5 \cdot 2$ in the first equation, $-26 + b = 128$

$$\Leftrightarrow \quad b = 154$$

The law is therefore $V = 154 - 5 \cdot 2t$

(*Check.* $t = 5$ gives $V = 154 - 26 = 128$)

Exercise 7

1   Find the equations of the straight lines shown in Figure 12 in the form $y = ax + b$ by substituting the coordinates of the given points in this equation.

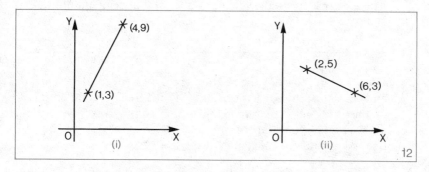

12

2   If $Q = at + b$, find the constants a and b for each of the following cases:

a $Q = 7$ when $t = 2$, and $Q = 13$ when $t = 6$
b $Q = 10$ when $t = 1$, and $Q = 1$ when $t = 4$
c $Q = 2 \cdot 1$ when $t = 2$, and $Q = 3 \cdot 3$ when $t = 5$
d $Q = 12 \cdot 5$ when $t = 10$, and $Q = 16 \cdot 5$ when $t = 5$.

3   Each of the following tables gives values of x and y which are connected by a law of the form $y = ax + b$. Plot the corresponding points and draw the line of best fit for the data. Hence determine the constants a and b in each case.

a

x	1	2	3	4	5	6
y	2·1	2·6	3·3	3·9	4·6	5·1

b

x	5	10	15	20	25	30
y	43	36	30	22	14	8

4   In an experiment, the data relating two quantities P and L are shown in the following table:

L	2	4	6	8	10
P	0·52	0·64	0·76	0·88	1·00

a Show this information on a graph, taking 1 cm to 1 unit (horizontally) on the L-axis, 1 cm to 0·1 unit on the P-axis, and origin at the point (2, 0·50).

b Draw the best-fitting line for the points plotted.

c Using the graph, and assuming that P and L are connected by a law of the form $P = aL + b$, determine the constants a and b, and hence state the law.

5   The electrical resistance R ohms of a wire at temperature t °C is shown below:

t	10	20	40	60	80	100
R	11·3	11·7	12·5	13·3	14·1	14·9

a Draw a graph to show the relation given by the formula $R = at + b$, and determine the constants a and b. (Take 1 cm to represent 10 °C on the t-axis and 2 cm to represent 1 ohm on the R-axis, with origin at the point (0, 10).) State the formula.

b Use the formula to calculate R when $t = 17·5$.

6   The speed v metres per second of a train t seconds after the brakes are applied is as follows:

t	0	5	10	15	20	25	30
v	30	25	22	18	14	11	6

a Plot the ordered pairs (t, v) using a scale of 2 cm to 5 units on both axes, and draw the best-fitting line.

b Find a law in the form $v = at + b$, where a and b are constants.

c Calculate (1) v when $t = 8$ (2) t when $v = 0$.

7   The molar volume V of a certain hydrocarbon series is related to the number of carbon atoms x as follows:

x	1	2	3	4	5
V	64	88	112	136	160

It is suspected that a law of the form $V = ax + b$ exists. Find if this is the case, and if so calculate a and b, and give the formula.

8   The volume of water Q litres in a storage tank after t seconds, when water is flowing freely from an outlet at the base, is given in the following table:

t	5	10	15	20	25
Q	487	475	462	450	437

Plot these values to suitable scales and draw the best-fitting line. Hence find a formula connecting Q and t in the form $Q = at + b$. Deduce from your formula the approximate time taken, in minutes and seconds, to empty the tank.

Summary

1   *Direct variation*

a y varies (directly) as x, or $y \propto x$

$\Leftrightarrow \quad \dfrac{y}{x} = k, \quad$ or $\quad y = kx.$

b y varies (directly) as x^n, or $y \propto x^n$

$\Leftrightarrow \quad \dfrac{y}{x^n} = k, \quad$ or $\quad y = kx^n,$

where k is the constant of variation.

2   *Inverse variation*

a y varies inversely as x, or $y \propto \dfrac{1}{x}$

$\Leftrightarrow \quad xy = k, \quad$ or $\quad y = \dfrac{k}{x}.$

b y varies inversely as x^n, or $y \propto \dfrac{1}{x^n}$

$\Leftrightarrow \quad x^n y = k, \quad$ or $\quad y = \dfrac{k}{x^n}.$

3   *Joint variation*

a y varies jointly as x and z, or $y \propto xz$

$\Leftrightarrow \quad y = kxz.$

b F varies jointly as e_1 and e_2 and inversely as r^2

$\Leftrightarrow \quad F = \dfrac{ke_1 e_2}{r^2}$

4   *Linear law of variation*

If x and y are variables whose replacements give points that lie on a straight line, then the formula, or law, connecting x and y is of the form: $y = ax + b$. In the language of variation, y varies partly as x and is partly constant.

Linear laws from experimental data may be found by first drawing a best-fitting line through the points. Choose two well-separated points (x_1, y_1) and (x_2, y_2) on this line; the equations $y_1 = ax_1 + b$ and $y_2 = ax_2 + b$ can then be solved for a and b.

3

Quadratic Equations and Inequations

1 *The open sentence* $ab = 0$

In this chapter we assume that the variables are on the set of real numbers.

Exercise 1

1 *a* What replacement for p makes $2p = 0$ a true sentence?
   b What replacement for q makes $5q = 0$ a true sentence?

2   What replacements for the variables in the following give true sentences?

   a $2(x-1) = 0$ *b* $4(t-5) = 0$ *c* $5(p+4) = 0$

   d $\frac{1}{2}(x+3) = 0$ *e* $3(2z-1) = 0$ *f* $7(3y+2) = 0$

3 *a* If $b \neq 0$, what replacement for a makes $ab = 0$ a true sentence?
   b If $a \neq 0$, what replacement for b makes $ab = 0$ a true sentence?
   c Is it true that $ab = 0 \Rightarrow a = 0$ *or* $b = 0$ *or* both a and b equal zero?

4   What replacements for the variables in the following give true sentences?

   a $(x-1)(x-2) = 0$ *b* $(y-5)(y+5) = 0$ *c* $(z+1)(2z-1) = 0$

   d $(2t-3)(3t-2) = 0$ *e* $c(c+7) = 0$ *f* $2x(x-9) = 0$

   g $3y(2y-1) = 0$ *h* $3x^2 = 0$ *i* $3(z-1)(z+5) = 0$

53

Write down the solution set of each of the following equations:

5   $(x+2)(x-4)=0$ 6 $(p-3)(p-8)=0$

7   $(y-3)(2y+1)=0$ 8 $(q+10)(q-10)=0$

9   $(2x-1)(2x+3)=0$ 10 $(3w+1)(3w-1)=0$

11   $r(r+4)=0$ 12 $x(x-10)=0$

13   $c(5c-2)=0$ 14 $3s(3s-1)=0$

15   $4x(7-x)=0$ 16 $2y(3+2y)=0$

17   $(x-\frac{1}{2})(x-\frac{2}{3})=0$ 18 $(2y-4)(5y+1)=0$

19   $2(z-6)(3z+9)=0$ 20 $8(5-2t)(4-2t)=0$

21   $(z-7)(z+7)=0$ 22 $(2r-5)^2=0$

23   $(x-1)(x-2)(x-3)=0$ 24 $(x+2)(2x-1)(3-x)=0$

---

2 Solving quadratic equations by factorisation

The equation $x^2+2x-3=0$, which contains x^2 and no higher powers of x, is called a *quadratic equation* or an *equation of the second degree* in x.

To *solve* this equation, we have to find all the replacements for x which satisfy $x^2+2x-3=0$, i.e. make the sentence true. These replacements are the *solutions* or *roots* of the quadratic equation.

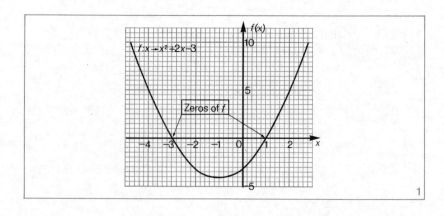

1

Since the solutions of $x^2 + 2x - 3 = 0$ correspond to the zeros of the quadratic function $f: x \to x^2 + 2x - 3$, we can solve the equation by seeing where the graph of f cuts the x-axis (which was explained in Chapter 3, Algebra, of Book 5). Figure 1 shows part of the graph, from which the zeros of f are -3 and 1. Hence the solution set of the quadratic equation $x^2 + 2x - 3 = 0$ is $\{-3, 1\}$.

The equation $ax^2 + bx + c = 0$, where $a \neq 0$, is called a *quadratic equation* or an *equation of the second degree* in x.

When $ax^2 + bx + c$ can be factorised, the solution set of the quadratic equation is quickly obtained without using a graph. The method uses the fact that if $p, q \in R$, then $pq = 0$ implies that $p = 0$, or $q = 0$ or both p and q equal zero. This is a property of the system of real numbers.

Example 1. Solve $9x^2 - 1 = 0$.

$$9x^2 - 1 = 0$$
$$\Leftrightarrow \quad (3x - 1)(3x + 1) = 0$$
$$\Leftrightarrow 3x - 1 = 0 \text{ or } 3x + 1 = 0$$
$$\Leftrightarrow \quad x = \tfrac{1}{3} \text{ or } x = -\tfrac{1}{3}$$

The solution set is $\{-\tfrac{1}{3}, \tfrac{1}{3}\}$.

Example 2. Solve $4x^2 - 4x + 1 = 0$.

$$4x^2 - 4x + 1 = 0$$
$$\Leftrightarrow \quad (2x - 1)(2x - 1) = 0$$
$$\Leftrightarrow 2x - 1 = 0 \text{ or } 2x - 1 = 0$$
$$\Leftrightarrow \quad x = \tfrac{1}{2} \text{ twice}$$

The solution set is $\{\tfrac{1}{2}\}$.

Note. In Example 2, we sometimes say that the quadratic equation has *equal roots*, or a *repeated root*.

Exercise 2

In questions *1–9*, solve each equation by first taking out the common factor:

1 $x^2 + 2x = 0$ *2* $x^2 - 5x = 0$ *3* $y^2 + 8y = 0$

4 $2z^2 - z = 0$ *5* $5x^2 - 2x = 0$ *6* $3x - 4x^2 = 0$

7 $2t^2 - 4t = 0$ *8* $7p^2 + 21p = 0$ *9* $6n - 2n^2 = 0$

In questions *10–18*, use the factors of a difference of two squares, $a^2 - b^2 = (a - b)(a + b)$, to solve:

10 $x^2 - 9 = 0$ *11* $4z^2 - 1 = 0$ *12* $9p^2 - 4 = 0$

13 $4m^2 - 49 = 0$ *14* $1 - y^2 = 0$ *15* $9 - 4t^2 = 0$

16 $\quad 16 - x^2 = 0$ *17* $\quad 25w^2 = 100$ *18* $\quad 36 = \frac{1}{4}x^2$

Find the solution set of each of the following quadratic equations:

19 $\quad x^2 = 0$ *20* $\quad 9x^2 = 0$ *21* $\quad (x-2)^2 = 0$

In questions *22–45*, factorise the quadratic expression on the left side of each equation, and hence solve:

22 $\quad x^2 - 3x + 2 = 0$ *23* $\quad y^2 - 7y + 12 = 0$ *24* $\quad z^2 - 6z + 5 = 0$

25 $\quad m^2 - m - 6 = 0$ *26* $\quad t^2 + 2t - 15 = 0$ *27* $\quad p^2 + 5p + 4 = 0$

28 $\quad x^2 + 10x + 21 = 0$ *29* $\quad a^2 + 4a + 3 = 0$ *30* $\quad x^2 - 6x + 9 = 0$

31 $\quad q^2 + 20q + 100 = 0$ *32* $\quad 2x^2 - 5x + 2 = 0$ *33* $\quad 2a^2 + 5a - 3 = 0$

34 $\quad 2a^2 - 5a + 3 = 0$ *35* $\quad 3t^2 - 10t + 3 = 0$ *36* $\quad 3y^2 - 8y - 3 = 0$

37 $\quad 5v^2 - v - 4 = 0$ *38* $\quad 2x^2 + 3x - 35 = 0$ *39* $\quad 12 - 19x + 4x^2 = 0$

40 $\quad 15 - 2y - y^2 = 0$ *41* $\quad 16 + 24d + 9d^2 = 0$ *42* $\quad 6 + 5w - 6w^2 = 0$

43 $\quad 2p^2 + 11p + 5 = 0$ *44* $\quad 9x^2 - 12x + 4 = 0$ *45* $\quad 12 + 7x - 12x^2 = 0$

Some quadratic equations are not in the standard form, $ax^2 + bx + c = 0$; for example

$$3x^2 = x + 10, \quad \text{and} \quad x + \frac{3}{x} = \frac{3 - 2x}{x}.$$

To check whether the first can be solved by factorisation, it must be expressed in the form $3x^2 - x - 10 = 0$.

In the second equation, each side may be cleared of fractions by multiplying throughout by a suitable non-zero number or expression. The most effective multiplier is the LCM of the denominators.

Example 1. Solve $3x^2 = x + 10$.

$$3x^2 = x + 10$$

$$\Leftrightarrow \quad 3x^2 - x - 10 = 0$$

$$\Leftrightarrow \quad (3x + 5)(x - 2) = 0$$

$$\Leftrightarrow 3x + 5 = 0 \text{ or } x - 2 = 0$$

$$\Leftrightarrow x = -\tfrac{5}{3} \text{ or } x = 2$$

The solution set is $\{-\tfrac{5}{3}, 2\}$

Example 2. Solve $x + \frac{3}{x} = \frac{3 - 2x}{x}$.

$$x + \frac{3}{x} = \frac{3 - 2x}{x} \qquad \text{(i)}$$

Multiplying each side by x,

$$x^2 + 3 = 3 - 2x$$

$$\Leftrightarrow \quad x^2 + 2x = 0 \qquad \text{(ii)}$$

$$\Leftrightarrow \quad x(x + 2) = 0$$

$$\Leftrightarrow x = 0 \text{ or } x = -2$$

Notice that in Example 2 we multiplied each side of the equation by x without knowing whether or not x can be replaced by zero. We therefore do not know if (i) and (ii) are equivalent equations unless we check that the roots of (ii) are also the roots of (i).

Check. Replacing x by 0 in (i), the left-side $= 0 + \frac{3}{0}$ which is meaningless, since division by zero is not possible. Without proceeding further, 0 is not a solution of (i).

Replacing x by -2 in (i),

$$\text{L.S.} = -2 - \frac{3}{2} = -\frac{7}{2} \text{ and R.S.} = \frac{3+4}{-2} = -\frac{7}{2}$$

and so -2 does check. The solution set of (i) is $\{-2\}$, which is a subset of the solution set $\{-2, 0\}$ of (ii).

Example 2 shows that when each side of an equation is multiplied by an expression containing the variable, the resulting equation and the original equation are not necessarily equivalent. As a consequence, the roots of the secondary equation should be checked for validity. In many cases, the check can be done mentally.

Exercise 2B

Solve the following by factorisation:

1 $2x^2 + 5x + 2 = 0$ *2* $4x^2 + 8x + 3 = 0$ *3* $10 + x - 2x^2 = 0$

4 $14 + 17t - 6t^2 = 0$ *5* $2y^2 + 5y - 12 = 0$ *6* $6z^2 - 13z + 5 = 0$

7 $6p^2 + 19p - 7 = 0$ *8* $21 - 8m - 4m^2 = 0$ *9* $9t^2 + 12t + 4 = 0$

10 $3x^2 + 12x + 12 = 0$ *11* $5v^2 - 30v + 45 = 0$ *12* $2w^2 + 6w - 56 = 0$

Solve the following by first rearranging each equation so that one side is zero:

13 $2x^2 + 5x = 7$ *14* $n^2 = 10n + 24$ *15* $6y^2 = 7y + 3$

16 $p(p+4) = 32$ *17* $t(t-5) = 24$ *18* $4x(x+1) = 15$

19 $(x-3)(2x+3) = 5$ *20* $(2a-3)^2 = 1$ *21* $(3b-1)^2 = 4$

22 $x^2 + (x-1)^2 = 1$ *23* $(3x-2)(x+1) = 1 + x$

24 $(v+1)(v-1) = 5(v+1)$ *25* $(x+2)(x+3) = x + 3$

26 $2(x-3) = (2x+3)(3-x)$ *27* $(x+1)^2 + 6(x+1) + 8 = 0$

28 $(2x-5)^2 + (2x-5) = 12$ *29* $3 + 7(x-3) = 6(x-3)^2$

30   Given that $x = 2$ is a root of $x^2 - 1 = a(2x - a - 8)$, find a.

Remove the fractions, and hence solve the following equations:

31 $\frac{1}{2}x^2 + \frac{3}{2}x = 9$ *32* $\frac{1}{2}y^2 + \frac{3}{4}y = 5$

33 $\frac{x}{2} - \frac{3}{x} = \frac{5}{2}$ *34* $x + \frac{2}{x} = 3$

35 $x + \frac{4}{x} = 4$ *36* $x - \frac{2}{x} = 1$

37 $x - \frac{9}{x} = 8$ *38* $2x - 5 = \frac{12}{x}$

39 $x - \frac{1}{x} = \frac{3x}{4}$ *40* $\frac{1}{2}x(x+1) = 15$

41 $\frac{1}{3}x(2x-3) = 9$ *42* $\frac{x}{4} - \frac{2x-1}{x+1} = 1$

43 $\frac{1}{x} + \frac{1}{x-1} = \frac{3}{2}$ *44* $\frac{1}{x} - \frac{1}{x+1} = \frac{1}{6}$

45 $\frac{2x-1}{2x+1} = \frac{x+2}{3x-2}$ *46* $\frac{3}{x+1} + \frac{1}{x-1} = 2$

47 $\frac{1}{x-2} - \frac{1}{x+3} = \frac{1}{10}$ *48* $\frac{x-1}{x+3} + \frac{x-2}{x+1} = 1$

49 $\frac{4}{x-1} - \frac{3}{x+1} = \frac{1}{2}$ *50* $\frac{1}{(v+1)^2} = \frac{4}{v+1} + 2.$

---

3 *Problems involving quadratic equations*

Example 1. The area of the floor of a rectangular workshop has to be 144 m^2, and the length must be 10 m more than the breadth. Find the length and breadth of the workshop.

Let l m represent the length; then $(l-10)$ m represents the breadth.

We now have
$$l(l-10) = 144$$
$$\Leftrightarrow \quad l^2 - 10l - 144 = 0$$
$$\Leftrightarrow \quad (l+8)(l-18) = 0$$
Since $\quad l > 0, \; l+8 \neq 0$
so $\quad l-18 = 0$, i.e. $l = 18$

Thus the length is 18 m and the breadth is $(18-10)$ m, i.e. 8 metres.

Example 2. The sum S of the first n natural numbers is given by the formula $S = \frac{1}{2}n(n+1)$. How many consecutive natural numbers, starting at 1, must be added to give 66?

Let n be the required number.

$$\frac{1}{2}n(n+1) = 66$$
$$\Leftrightarrow \quad n^2 + n = 132$$
$$\Leftrightarrow \quad n^2 + n - 132 = 0$$
$$\Leftrightarrow \quad (n+12)(n-11) = 0$$
Since $\quad n > 0, \quad n+12 \neq 0$
so $\quad n-11 = 0$, i.e. $n = 11$

So the first 11 natural numbers add up to 66.

Exercise 3

1   Two whole numbers differ by 7 and their product is 78. Let the numbers be n and $n+7$. Form a quadratic equation in n and hence obtain the two numbers.

2   Two whole numbers differ by 2 and their product is 168. What are the numbers?

3   The sum of two whole numbers is 25 and their product is 136.

a   If n is one of the numbers, express the other number in terms of n.
b   Form a quadratic equation in n and solve it. Hence state the numbers.

4   Figure 2 shows the plan of a rectangular floor of area 180 m². The breadth of the floor is 3 metres less than the length. Use the given information to find the length and breadth of the floor.

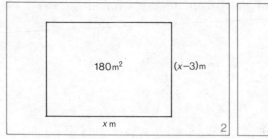

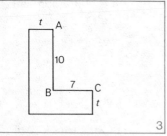

5   Figure 3 shows an L-shaped piece of metal of uniform width t cm with AB = 10 cm and BC = 7 cm. Show that the area A cm^2 of the metal is given by the formula $A = t^2 + 17t$. Find t when $A = 60$.

6   The area of a sheet of drawing paper is 300 cm^2. The length is 5 cm more than the breadth. Find the length and breadth of the paper.

7   The area of a rectangular pane of glass is 1500 cm^2, and its perimeter is 160 cm.

   a Let x cm be the length of the pane. Write down the breadth of the pane in terms of x.
   b Form an equation in x and solve it. Hence state the dimensions of the pane.

8   A stone is thrown vertically upwards at a speed of 24 m/s. Its height h metres after t seconds is given approximately by the formula $h = 24t - 5t^2$. Use this formula to find when the stone is 27 metres up, and explain the double answer.

9   The perimeter of a rectangular plot of ground is 42 m and its area is 80 m^2. Find the length and breadth of the plot.

10   The height of a triangle is 5 cm more than the base. If the area of the triangle is 75 cm^2, find its height.

11   The sum S of the first n natural numbers is given by the formula $S = \frac{1}{2} n(n+1)$. How many consecutive natural numbers, starting at 1, must be added together to give 210?

12   The sum S of the first n even numbers, starting at 0, is given by the formula $S = n(n-1)$. How many consecutive even numbers, starting at 0, add up to 156?

13   Figure 4 shows part of the parabola $y = 6x - x^2$.

a Given that the point $(3, k)$ lies on the parabola, find k.

b If MP = 8, find OM.

c Find the coordinates of C.

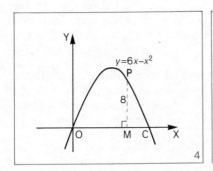

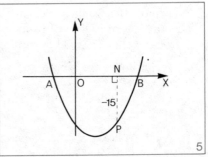

14 The curve in Figure 5, whose equation is $y = x^2 - 4x - 12$, cuts the x-axis in the points A and B.

a Find the coordinates of A and B.

b Find the coordinates of P when NP = -15.

15 The sides of a right-angled triangle are n, $n+1$, $n+2$ cm long. Which side is the hypotenuse? Find n by solving an equation, and state the area of the triangle.

Exercise 3B

1 The sum of a number n and its reciprocal is 2. Form an equation and hence find n.

2 The sum of a number and its reciprocal is $\frac{29}{10}$. Find the number.

3 The equation of a curve is $y = 2x + 5/x$. The point $(a, 11)$ lies on the curve. Find two possible values of a.

4 A curve has equation $x^2 + y^2 + 7x - 8y + 12 = 0$. Find the coordinates of the points where the curve cuts: *a* the x-axis *b* the y-axis.

5 The formula $M = \frac{1}{2}wx(L - x)$ occurs in the theory of the bending of beams. Find x when $M = 2700$, $w = 200$ and $L = 12$.

6 A thin circular washer has an outer radius of x cm and a width of t cm.

a Write down the inner radius in terms of x and t.

    b Show that the area A cm^2 of one of its plane surfaces is given by the formula $A = \pi t(2x - t)$.

    c If $A = 66$, $x = 5$ and $\frac{22}{7}$ is taken as an approximation for π, use the formula to find t.

7   From a sheet of paper 12 cm long and 9 cm wide, a strip of uniform width x cm is trimmed off from each of the four sides so that the area of the rectangle remaining is 54 cm^2. Illustrate by a diagram.

    a Write down the length and breadth of the smaller sheet in terms of x.

    b Form a quadratic equation in x, and solve it to find the width of the strips cut off.

8   A lidless box is in the shape of a cube of external side x cm and is made of wood 1 cm thick.

    a Write down the internal length, breadth and height of the box in terms of x.

    b Show that the volume of wood, V cm^3, forming the box is given by the formula $V = 5x^2 - 8x + 4$.

    c Find x when $V = 424$.

9   A motor boat has a cruising speed in still water of 15 km/h. It takes a total time of 3 hours to go up-river for a distance of 20 km and back the same distance, the river current being steady.

    a Let the current be x km/h. Write down the time taken to go up-river and the time taken for the return journey, in terms of x.

    b Form an equation in x and solve it to find the speed of the current.

10   In triangle PQR, which is right-angled at Q, QR is 1 cm less than PQ and PR is 8 cm greater than PQ.

    a If PQ is x cm long, express QR and PR in terms of x.

    b Calculate the lengths of the three sides of triangle PQR.

11   Figure 6 shows the cross-section of a cylindrical oil tank of radius 2·5 m. The width AB of the exposed oil surface is $2x$ metres and the depth of oil at the midpoint E of AB is d metres.

    a Write down the length of CE in terms of d.

    b Apply Pythagoras' theorem to $\triangle$AEC to show that $x^2 - 5d + d^2 = 0$.

    c Find d when $x = 1\frac{1}{2}$.

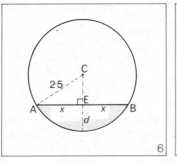

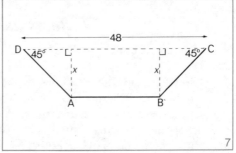

6

7

12 Figure 7 shows the cross-section of a cutting whose sloping sides AD and BC make an angle of 45° with the ground whose natural surface DC is horizontal.

a Show that the area of the cross-section is $x(48 - x)$ m².
b Find x for which this area is 320 m².

13 Figure 8 shows the graphs of $y = 2x - 5$ and $y = 8(x^2 - 1)$.

a Show that at the points of intersection, $8x^2 - 2x - 3 = 0$.
b Hence find the coordinates of the points of intersection A and B.

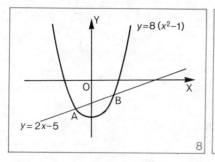

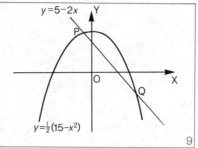

8

9

14 Figure 9 shows the line $y = 5 - 2x$ and the parabola $y = \frac{1}{2}(15 - x^2)$. Find x at the points of intersection P and Q, and hence state the coordinates of these points.

4 Roots and equations

Solving quadratic equations by the method of factors, as discussed in Section 2, is not always suitable or practicable. We therefore investigate an alternative method which depends on the fact that every positive real number has two real square roots which are additive inverses of each other.

Example 1. Solve $x^2 - 9 = 0$.

$$x^2 - 9 = 0 \Leftrightarrow x^2 = 9$$

Since $3^2 = 9$ and $(-3)^2 = 9$, 3 and -3 are the square roots of 9.

The solution set is $\{-3, 3\}$.

Layout of working

$$x^2 - 9 = 0$$

$$\Leftrightarrow \quad x^2 = 9$$

$$\Leftrightarrow \quad x = \pm\sqrt{9}, \text{ i.e. } x = \pm 3$$

The solution set is $\{-3, 3\}$.

Example 2. Solve $x^2 - 2 = 0$.

$$x^2 - 2 = 0$$

$$\Leftrightarrow \quad x^2 = 2$$

$$\Leftrightarrow \quad x = \pm\sqrt{2}$$

Since $\sqrt{2}$ and $-\sqrt{2}$ are irrational numbers (as explained in Chapter 1 of Book 4), the equation is said to have irrational roots.

The solution set is $\{-\sqrt{2}, \sqrt{2}\}$.

Example 3. Solve $(2x - 1)^2 = 25$.

$$(2x - 1)^2 = 25$$

$$\Leftrightarrow \quad 2x - 1 = \pm 5$$

$$\Leftrightarrow \quad 2x = 1 \pm 5$$

$$\Leftrightarrow \quad 2x = 6 \text{ or } -4$$

$$\Leftrightarrow \quad x = 3 \text{ or } -2$$

The solution set is $\{-2, 3\}$.

Note. To avoid ambiguity in the meaning of the symbol $\sqrt{}$, $\sqrt{9}$ is defined to be the *positive* square root of 9; the *negative* square root of 9 is denoted by $-\sqrt{9}$. Similarly, $\sqrt{2}$ and $-\sqrt{2}$ are the positive and negative square roots of 2, and so on for all positive real numbers. Zero has only one square root, namely zero itself.

The equation $x^2 + 4 = 0$ leads to a very different situation since it implies that $x^2 = -4$. Since the square of every real number is positive or zero no real number satisfies this equation. The solution set is ϕ.

Exercise 4B

By taking the square root of each side of the equations in questions *1–8*, find the solution sets.

1 $x^2 = 36$ *2* $x^2 = 100$ *3* $x^2 + 1 = 0$ *4* $x^2 = 3$

5 $x^2 = \frac{1}{4}$ *6* $x^2 = \frac{9}{16}$ *7* $4x^2 - 1 = 0$ *8* $4x^2 - 25 = 0$

Using the square-root method shown in Worked Example 3, solve each of these:

9 $(x-1)^2 = 4$ *10* $(x+1)^2 = 4$ *11* $(x-3)^2 = 100$

12 $(x-2)^2 = 16$ *13* $(x+5)^2 = 9$ *14* $(x+9)^2 = 25$

15 $(x-2)^2 - 16 = 0$ *16* $(x-\frac{1}{2})^2 = 25$ *17* $(x+\frac{1}{2})^2 = 1$

18 $(x+\frac{1}{4})^2 = 9$ *19* $(x-\frac{3}{4})^2 - 36 = 0$ *20* $(2x-1)^2 = 81$

21 $(3x-2)^2 - 49 = 0$ *22* $(5-2x)^2 = 25$ *23* $(5-3x)^2 = 1$

In questions *24–32*, solve for x in terms of the constants and other variables.

24 $x^2 = 9y^2$ *25* $x^2 = 4y, \ y > 0$ *26* $y = \frac{1}{2}x^2, \ y > 0$

27 $y^2 = 4x^2$ *28* $x^2 + y^2 = a^2$ *29* $a^2 - x^2 = b^2$

30 $x^2 - p^2 = q^2 - 3x^2$ *31* $a^2x^2 - b^2 = c^2x^2$ *32* $\dfrac{a^2}{x^2+1} = \dfrac{1}{x^2}$

Change the subject of each of the following formulae to the symbol indicated, taking the positive square root in each case.

33 $A = \pi r^2$ to r *34* $V = \pi r^2 h$ to r *35* $T = \frac{1}{2}mv^2$ to v

36 $V = \frac{1}{3}\pi r^2 h$ to r *37* $F = \dfrac{mv^2}{r}$ to v *38* $R = \dfrac{kL}{D^2}$ to D

39 $A = \pi R^2 - \pi r^2$ to: $a \ R \ b \ r$ *40* $V = \pi h(R^2 - r^2)$ to: $a \ R \ b \ r$

41 $b^2 = a^2(1 - e^2)$ to e *42* $T = c\sqrt{(1+k^2)}$ to k

5 *Solving quadratic equations by completing the square*

9, x^2, $4x^2$, $(x-3)^2$, $(x+p)^2$ are examples of *perfect squares*. If a quadratic equation can be expressed in the form:

$$(x+p)^2 = q, \quad q > 0,$$

it may be solved by the method of Section 4. To extend the method, use will be made of the identity:

$$(x+p)^2 = x^2 + 2px + p^2$$

Example 1. Solve $x^2 - 6x + 1 = 0$.
$$x^2 - 6x + 1 = 0$$
$$\Leftrightarrow \quad x^2 - 6x = -1$$

Add ($\frac{1}{2}$ coefficient of x)^2 to each side of this equation to make the left side a perfect square.

$$x^2 - 6x + (-3)^2 = -1 + (-3)^2$$
$$\Leftrightarrow \qquad (x-3)^2 = 8$$
$$\Leftrightarrow \qquad x - 3 = \pm\sqrt{8}$$
$$\Leftrightarrow \qquad x = 3 \pm \sqrt{8}$$
$$\Rightarrow \quad x \doteqdot 3 + 2\cdot83 \ or \ x \doteqdot 3 - 2\cdot83$$
$$\Rightarrow \quad x \doteqdot 5\cdot83 \ or \ x \doteqdot 0\cdot17$$

The process of adding a constant term to a quadratic expression to make it a perfect square is called *completing the square*. In applying the process, we must ensure that the coefficient of x^2 is unity.

Example 2. Solve $2x^2 + 6x - 1 = 0$.

$$2x^2 + 6x - 1 = 0$$

$$\Leftrightarrow \quad x^2 + 3x \quad = \tfrac{1}{2}$$

$$\Leftrightarrow x^2 + 3x + (\tfrac{3}{2})^2 = \tfrac{1}{2} + (\tfrac{3}{2})^2$$

$$\Leftrightarrow \quad (x + \tfrac{3}{2})^2 = \tfrac{1}{2} + \tfrac{9}{4}$$

$$\Leftrightarrow \quad (x + 1 \cdot 5)^2 = 2 \cdot 75$$

$$\Rightarrow \quad x + 1 \cdot 5 \doteqdot \pm 1 \cdot 66$$

$$\Rightarrow \quad x \quad \doteqdot -1 \cdot 50 \pm 1 \cdot 66$$

$$\Rightarrow \quad x \doteqdot -3 \cdot 16 \text{ or } x \doteqdot 0 \cdot 16$$

To complete the square of $x^2 + 2px$, add ($\tfrac{1}{2}$ coefficient of $x)^2$, i.e. p^2, to give $x^2 + 2px + p^2 = (x + p)^2$.

Exercise 5B

In questions *1—12*, what number must be added to each to make a perfect square?

1	$x^2 + 4x$	*2*	$t^2 + 8t$	*3*	$x^2 + 10x$	*4*	$y^2 - 6y$
5	$n^2 - 2n$	*6*	$x^2 - 12x$	*7*	$x^2 + x$	*8*	$x^2 + 3x$
9	$x^2 + \tfrac{2}{3}x$	*10*	$x^2 - 7x$	*11*	$x^2 - 9x$	*12*	$z^2 - \tfrac{1}{2}z$

Complete the square in questions *13–20*.

13 $x^2 + 4x + \ldots = (x + \ldots)^2$ *14* $y^2 - 6y + \ldots = (y - \ldots)^2$

15 $t^2 + 12t + \ldots = (t + \ldots)^2$ *16* $x^2 - 14x + \ldots = (x - \ldots)^2$

17 $a^2 + a + \ldots = (a + \ldots)^2$ *18* $m^2 - m + \ldots = (m - \ldots)^2$

19 $z^2 - 0 \cdot 6z + \ldots = (z - \ldots)^2$ *20* $x^2 + 5x + \ldots = (x + \ldots)^2$

Solve the following quadratic equations by completing the square:

21 $x^2 + 2x = 8$ *22* $x^2 - 2x = 3$ *23* $x^2 - 4x = 21$

24 $x^2 - 4x + 3 = 0$ *25* $x^2 + 6x - 27 = 0$ *26* $x^2 - 6x - 16 = 0$

27 $x^2 + 10x - 24 = 0$ *28* $x^2 - 14x + 48 = 0$ *29* $x^2 + 10x + 25 = 0$

30 $x^2 + x - 2 = 0$ *31* $x^2 + 2x + 4 = 0$ *32* $x^2 + 5x + 6 = 0$

33 $x^2 + 3x - 40 = 0$ *34* $2x^2 + x - 6 = 0$ *35* $2x^2 + 3x - 14 = 0$

Solve the following by completing the square, rounding off the roots to 1 decimal place:

36 $x^2 + 4x + 1 = 0$ *37* $y^2 - 6y + 3 = 0$ *38* $t^2 + 6t + 4 = 0$

39 $z^2 - 2z - 1 = 0$ *40* $x^2 + 8x - 1 = 0$ *41* $x^2 + 3x + 1 = 0$

42 $t^2 - 3t - 2 = 0$ *43* $2x^2 + 2x - 1 = 0$ *44* $3m^2 + 6m - 2 = 0$

45 $5y^2 - 30y - 18 = 0$ *46* $2z^2 = 3(4z + 5)$ *47* $2x^2 = 3x + 6$

48 $4t^2 - 6t = 3$ *49* $-3x^2 + 9x + 14 = 0$ *50* $\dfrac{h^2 + 5}{2h} = 15$

6 Solving quadratic equations by formula

Every quadratic equation can be expressed in the *standard form* $ax^2 + bx + c = 0$, where $a \neq 0$. If we can solve this equation, we shall obtain a formula which can be used to find the solution set of any quadratic equation.

$$ax^2 + bx + c = 0$$

$$\Leftrightarrow \quad x^2 + \frac{b}{a}x = -\frac{c}{a}, \text{ since } a \neq 0$$

$$\Leftrightarrow x^2 + \frac{b}{a}x + \left(\frac{b}{2a}\right)^2 = \left(\frac{b}{2a}\right)^2 - \frac{c}{a}$$

$$\Leftrightarrow x^2 + \frac{b}{a}x + \left(\frac{b}{2a}\right)^2 = \frac{b^2}{4a^2} - \frac{c}{a}$$

$$\Leftrightarrow \quad \left(x + \frac{b}{2a}\right)^2 = \frac{b^2 - 4ac}{4a^2}$$

$$\Leftrightarrow \quad x + \frac{b}{2a} = \frac{\pm\sqrt{(b^2 - 4ac)}}{2a}$$

$$\Leftrightarrow \quad x = \frac{-b \pm \sqrt{(b^2 - 4ac)}}{2a}$$

This result is known as the *quadratic formula*. Before using it, the quadratic equation must be expressed in standard form.

Example 1. Solve $2x^2 + 3x = 4$.

$$2x^2 + 3x = 4$$
$$\Leftrightarrow 2x^2 + 3x - 4 = 0$$

Compare $ax^2 + bx + c = 0$

$a = 2, b = 3, c = -4$

$$x = \frac{-b \pm \sqrt{(b^2 - 4ac)}}{2a}$$

$$= \frac{-3 \pm \sqrt{[3^2 - 4.2.(-4)]}}{4}$$

$$= \frac{-3 \pm \sqrt{(9 + 32)}}{4}$$

$$= \frac{-3 \pm \sqrt{41}}{4}$$

$$\doteqdot \frac{-3 \pm 6 \cdot 40}{4}$$

$$= \frac{-3 + 6 \cdot 40}{4} \quad or \quad \frac{-3 - 6 \cdot 40}{4}$$

$$= \frac{3 \cdot 40}{4} \quad or \quad \frac{-9 \cdot 40}{4}$$

$$\doteqdot 0 \cdot 85 \quad or \quad -2 \cdot 35$$

$$= 0 \cdot 8 \quad or \quad -2 \cdot 4 \text{ to 1 place of}$$

decimals.

Example 2. Solve $2x^2 - 3x + 4 = 0$.

$$2x^2 - 3x + 4 = 0$$

Compare $ax^2 + bx + c = 0$

$a = 2, b = -3, c = 4$

$$x = \frac{-b \pm \sqrt{(b^2 - 4ac)}}{2a}$$

$$= \frac{3 \pm \sqrt{[(-3)^2 - 4.2.4]}}{4}$$

$$= \frac{3 \pm \sqrt{(9 - 32)}}{4}$$

$$= \frac{3 \pm \sqrt{(-23)}}{4}$$

$$\sqrt{(-23)} \notin R,$$

so the solution set is $\varnothing$.

Reminders: (i) $\sqrt{657} = \sqrt{(6 \cdot 57 \times 100)} = 2 \cdot 56 \times 10 = 25 \cdot 6$

(ii) Rounding-off procedure, as in Arithmetic, Chapter 1, of Book 2: If the next figure is greater than 5, increase the round-off figure by 1; if the following figure is 5, round-off to the nearest even number; otherwise leave the round-off figure as it is.

Exercise 6

Use the quadratic formula to solve the following equations:

1 $x^2 + 5x + 6 = 0$ *2* $x^2 - 3x + 2 = 0$ *3* $2x^2 + 7x + 3 = 0$

4 $2x^2 + 3x - 9 = 0$ *5* $2x^2 - 11x + 12 = 0$ *6* $3x^2 - 2x - 8 = 0$

7 $3t^2 + 10t - 8 = 0$ *8* $4y^2 - 8y - 5 = 0$ *9* $4m^2 - 20m + 25 = 0$

10 $3x^2 + 13x - 10 = 0$ *11* $3z^2 + 20z = 7$ *12* $4x^2 = 5(3x + 5)$

Solve the following quadratic equations by the quadratic formula, rounding the roots·off to one decimal place.

13 $x^2 + 4x + 1 = 0$ *14* $x^2 + 6x + 4 = 0$ *15* $x^2 + 7x + 5 = 0$

16 $x^2 + 2x - 1 = 0$ *17* $x^2 - 6x + 3 = 0$ *18* $x^2 - 4x - 7 = 0$

19 $x^2 + 8x = 10$ *20* $x^2 = 12x + 5$ *21* $x^2 + 20x + 15 = 0$

22 $2x^2 + x - 4 = 0$ *23* $2x^2 - 3x - 4 = 0$ *24* $2x^2 + 12x + 9 = 0$

25 $4x^2 - 12x + 3 = 0$ *26* $3x^2 - 12x + 11 = 0$ *27* $5y^2 + 3y - 4 = 0$

28 $5x^2 = 12x + 8$ *29* $10t^2 = 7t + 1$ *30* $15x^2 + 2x - 4 = 0$

31 $3x^2 + 10 = 18x$ *32* $5x^2 + 12x - 5 = 0$ *33* $10x^2 + 12x - 9 = 0$

34 $x^2 - 10x - 15 = 0$ *35* $4x^2 = 3(4x + 5)$ *36* $2x^2 + 15x + 16 = 0$

37 $d^2 - 1\cdot3d - 1\cdot31 = 0$ *38* $x^2 - 1\cdot2x + 0\cdot09 = 0$

39 $x(x - 1) = 3(x + 1)$ *40* $(2x - 1)(3x + 2) - (x + 3)(2x + 1) = 0$

41 $x + \dfrac{1}{x} = 3, \ x \neq 0$ *42* $\dfrac{1}{x} + \dfrac{x}{3} = 2, \ x \neq 0$

43 Solve the following quadratic equations, expressing the roots in terms of a:

  a $x^2 + 5x + a = 0$ *b* $2x^2 - ax + 9 = 0$

  c $ax^2 + 2x + a = 0$ *d* $x^2 + 4x + (5 - a) = 0$

44 Find a quadratic equation which is equivalent to the equation

$$1\cdot4x^2 - 2\cdot8x = 1\cdot8$$

but has integral coefficients. Hence find the roots of the given equation to one decimal place.

7 Graphs of subsets of real numbers: intervals

Subsets of R such as $\{x: -3 < x < 5\}$ are often called *intervals*. The *graph* of an interval on the real number line is a well-defined line segment which is shown as a heavy black line suitably indicated. If the beginning (or the end) of the segment is *included*, a black dot is marked at the beginning (or the end) of the segment; but if it is *not included*, an open circle is marked. The examples given below illustrate the method. Look at Figure 10.

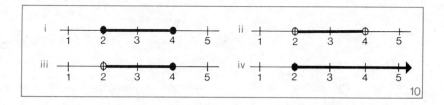

10

(i)   shows the graph of the *closed interval* $\{x: 2 \leqslant x \leqslant 4\}$. We can read this as 'the closed interval two, four.' 2 and 4 are the *end points* of the interval.

(ii) shows the graph of the *open interval* $\{x: 2 < x < 4\}$. We can read this as 'the open interval two, four'.

(iii) shows the graph of the interval $\{x: 2 < x \leqslant 4\}$. In this interval, 4 is included but 2 is excluded. We say that the interval is *open* at 2 and *closed* at 4.

(iv) shows the graph of the interval $\{x: x \geqslant 2\}$. The arrowhead indicates that the graph continues indefinitely in the direction shown. Since intervals are sets, we can have intersection and union of intervals.

Exercise 7

Using set-builder notation, write down the intervals whose graphs are shown in questions *1–6*:

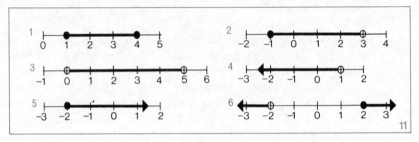

In questions *7–15*, draw the graphs of the following intervals:

7   $\{x: x > 3\}$ 8 $\{x: x \geqslant 1\}$ 9 $\{x: x \leqslant -2\}$

10   $\{x: x > -1\}$ 11 $\{x: -1 \leqslant x \leqslant 3\}$ 12 $\{x: x < -2 \text{ or } x \geqslant 2\}$

13   $\{x: x > -2 \text{ and } x < 4\}$ 14 $\{x: 0 \leqslant x \leqslant 5\}$ 15 $\{x: x \leqslant 0 \text{ or } x > 3\}$

In questions *16–23*, draw a graph to show the solution set of each of the following:

16   $x - 1 < 3$ 17 $x + 1 > 0$ 18 $x - 1 \leqslant 0$ 19 $-3 < x < 3$

20   $2x \leqslant 7$ 21 $2x - 1 > 5$ 22 $5 - x \geqslant 1$ 23 $\frac{1}{3}(1 - x) < 1$

24   Draw the graphs of the following on number lines below one another.

     a $P = \{x: x \geqslant 2, x \in R\}$ *b* $Q = \{x: x \leqslant -1, x \in R\}$ *c* $P \cup Q$

25   On separate number lines draw the graphs of $S = \{x: x > -3\}$, $T = \{x: x < 2\}$, and $S \cap T$, where $x \in R$. Give a set-builder description of $S \cap T$.

8 Solving quadratic inequations

$2x^2 + 3x - 5 > 0$ and $2x^2 + 3x - 5 < 0$ are called *quadratic inequations* in *x*. The solution set of each may be found from the graph of the associated function $f: x \to 2x^2 + 3x - 5$, which is the *parabola* whose equation is $y = 2x^2 + 3x - 5$. The graph is shown in Figure 12.

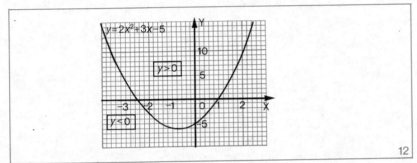

The x-coordinates of the points on the part of the graph which is *above* the x-axis give the solution set of the inequation $2x^2 + 3x - 5 > 0$. The x-coordinates of the points on the part *below* the x-axis give the solution set of the inequation $2x^2 + 3x - 5 < 0$.

The changes from positive to negative values of y occur at the points where the parabola crosses the x-axis. These points correspond to the zeros of f which are obtained by solving the quadratic equation:

$$2x^2 + 3x - 5 = 0$$
$$\Leftrightarrow (2x + 5)(x - 1) = 0$$
$$\Leftrightarrow x = -2\tfrac{1}{2} \text{ or } x = 1$$

Hence, the solution set of $2x^2 + 3x - 5 > 0$ is $\{x: x < -2\tfrac{1}{2} \text{ or } x > 1\}$, and the solution set of $2x^2 + 3x - 5 < 0$ is $\{x: x > -2\tfrac{1}{2} \text{ and } x < 1\}$, i.e. $\{x: -2\tfrac{1}{2} < x < 1\}$.

The graphs of the solution sets are shown in Figure 13.

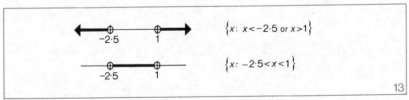

Example. Find the solution set of the inequation $12 + 4x - x^2 \geqslant 0$, and illustrate on the number line.

We first *sketch* the graph of $y = 12 + 4x - x^2$, which we know is a parabola.

If $y = 0$, $12 + 4x - x^2 = 0$
$\Leftrightarrow (2 + x)(6 - x) = 0$
$\Leftrightarrow x = -2$ or $x = 6$

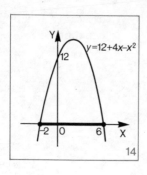

So the graph cuts the x-axis at $(-2, 0)$ and $(6, 0)$.
If $x = 0$, $y = 12$.
So the graph cuts the y-axis at $(0, 12)$.

From the sketch graph shown in Figure 14, the solution set of $12 + 4x - x^2 \geqslant 0$ is $\{x: -2 \leqslant x \leqslant 6\}$. The graph of the solution set is shown by the heavy line.

Note. The maximum turning point of the graph is easily found to be $(2, 16)$, if necessary.

Exercise 8

1 Figure 15 shows the graph of $y = x^2 - 4$. From the graph, write down the solution set of each of the following:

  a $x^2 - 4 = 0$ *b* $x^2 - 4 < 0$ *c* $x^2 - 4 > 0$

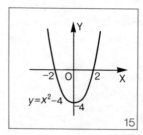
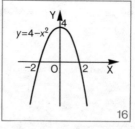
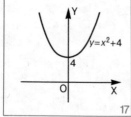

2 The graph of $y = 4 - x^2$ is shown in Figure 16. From the graph, find the solution set of each of the following:

  a $4 - x^2 = 0$ *b* $4 - x^2 \geqslant 0$ *c* $4 - x^2 < 0$.

3 Figure 17 shows the graph of $y = x^2 + 4$. From the graph, write down the solution set of:

  a $x^2 + 4 = 0$ *b* $x^2 + 4 > 0$ *c* $x^2 + 4 < 0$.

4 a If $y = (x+1)(x-3)$, write down the roots of $(x+1)(x-3) = 0$ and state the value of y when $x = 0$.

 b Use this information to sketch the graph of $y = (x+1)(x-3)$.

 c Hence find the solution set of:

  (1) $(x+1)(x-3) < 0$ *(2)* $(x+1)(x-3) > 0$.

5 a If $y = (2+x)(4-x)$, write down the roots of $(2+x)(4-x) = 0$ and also the value of y when $x = 0$.

 b Hence sketch the graph of $y = (2+x)(4-x)$ and use it to find the solution set of:

  (1) $(2+x)(4-x) \geqslant 0$ *(2)* $(2+x)(4-x) < 0$

6 If $y = 4x - x^2$, find x when $y = 0$ and also find y when $x = 2$. Use a sketch graph to obtain the solution set of:

 a $4x - x^2 > 0$ *b* $4x - x^2 \leqslant 0$

7 a Solve the equation $x^2 - x - 6 = 0$.

 b Use a graphical method to find the solution sets of:

  (1) $x^2 - x - 6 > 0$ *(2)* $x^2 - x - 6 < 0$.

Find the solution set of each of the following inequations, and illustrate on the number line as in Figure 13:

8 $(x-1)(x-3) < 0$ *9* $(x+1)(x+5) > 0$ *10* $(3x+2)(x-1) \leqslant 0$

11 $x^2 - 9 \leqslant 0$ *12* $x^2 + x - 12 > 0$ *13* $2 + x - x^2 > 0$

14 $2x^2 + 11x + 5 \geqslant 0$ *15* $2x^2 - x - 15 < 0$ *16* $12 - 5x - 2x^2 < 0$

17 $x^2 - 4x + 4 > 0$ *18* $x^2 + 4x + 4 < 0$ *19* $8 + 18x > 5x^2$

20 The height h metres of an object at time t seconds is given by the equation $h = 30t - 5t^2$.

 a Find t when $h = 0$.

 b Show that h cannot exceed 45.

 c Find the interval for t such that $h > 20$.

Summary

1 The equation $ax^2 + bx + c = 0$, where $a \neq 0$, is a *quadratic equation*, or an equation of the second degree in x. A *root* of an equation is any member of its solution set.

2 *Methods of solving a quadratic equation:*

  (i) *By drawing the graph* of $y = ax^2 + bx + c$. The roots of $ax^2 + bx + c = 0$ are obtained by reading the x-coordinates of the points of intersection with the x-axis.

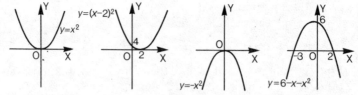

  (ii) *By factorisation.* Rearrange terms so that right side is zero.

  Example $5x^2 = 3x + 2$

  $\Leftrightarrow 5x^2 - 3x - 2 = 0$

  $\Leftrightarrow (5x + 2)(x - 1) = 0$

  $\Leftrightarrow x = -\frac{2}{5}$ or $x = 1$

  The solution set is $\{-\frac{2}{5}, 1\}$.

(iii) *By completing the square.* Reduce to form $(x+p)^2 = q$.

Example of (iii) above   *Example of (iv) below*

$2x^2 + 4x - 5 = 0$

$2x^2 + 4x = 5$

$x^2 + 2x = 2\cdot5$

$x^2 + 2x + 1^2 = 2\cdot5 + 1^2$

$(x+1)^2 = 3\cdot5$

$x + 1 = \pm\sqrt{3\cdot5}$

$x \doteqdot -1 \pm 1\cdot87$

$\doteqdot -2\cdot9 \text{ or } 0\cdot9$

to 1 decimal place.

$2x^2 + 4x - 5 = 0$

Here $a = 2$, $b = 4$, $c = -5$

$x = \dfrac{-4 \pm \sqrt{(16+40)}}{4}$

$= \dfrac{-4 \pm \sqrt{56}}{4}$

$\doteqdot \dfrac{-4 \pm 7\cdot48}{4}$

$= \dfrac{-4 - 7\cdot48}{4} \text{ or } \dfrac{-4 + 7\cdot48}{4}$

$= \dfrac{-11\cdot48}{4} \text{ or } \dfrac{3\cdot48}{4}$

$\doteqdot -2\cdot9 \text{ or } 0\cdot9$

(iv)   *By the quadratic formula*:

$$x = \frac{-b \pm \sqrt{(b^2 - 4ac)}}{2a}$$

3   *Quadratic inequations*: $ax^2 + bx + c > 0$ and $ax^2 + bx + c < 0$. Sketch the graph of $y = ax^2 + bx + c$ by finding its points of intersection with the x- and y-axes. Hence write down the solution sets of $ax^2 + bx + c > 0$ and $ax^2 + bx + c < 0$.

Revision Exercises

Revision Exercises on Chapter 1
Factors and Fractions

Revision Exercise 1A

Factorise the expressions in questions *1–24*:

1	$2x^2 + 6$	*2*	$ab^2x + a^2by$	*3* $p^2 + pq + pr$		*4*	$a^3 - a^2$
5	$g^2 - h^2$	*6*	$a^2 - 1$	*7* $b^2 - 100$		*8*	$2x^2 - 50$
9	$9a^2 - 25b^2$	*10*	$3 - 12y^2$	*11* $x^3 - x$		*12*	$1 - 49z^2$
13	$x^2 + x - 2$	*14*	$a^2 + 15a + 26$	*15* $p^2 + 4p - 5$			
16	$y^2 - 12y + 36$	*17*	$4x^2 + 16x + 15$	*18* $2a^2 - 9a + 4$			
19	$6x^2 + 13x + 6$	*20*	$6x^2 + 13x + 7$	*21* $6x^2 - 19x + 15$			
22	$3b^2 - 2b - 1$	*23*	$5a^2 - 16a + 3$	*24* $2x^2 - 5x - 3$			

25 △ABC is right-angled at A. Calculate c when:

   a $a = 2\cdot5, b = 1\cdot5$ *b* $a = 12\cdot5, b = 3\cdot5$ *c* $a = 20\cdot5, b = 4\cdot5$

26 Factorise the right-hand side of the formula $A = \pi(x^2 - y^2)$ to calculate A when 3·14 is an approximation for π and:

   a $x = 6\cdot6, y = 3\cdot4$ *b* $x = 85, y = 15$ *c* $x = 4\cdot2, y = 0\cdot8$

27 Use factors to calculate the value of $\frac{1}{2}mv^2 - \frac{1}{2}mu^2$ when $m = 5, v = 25$ and $u = 15$.

Simplify the expressions in questions *28–39*:

28 $\dfrac{x^2 - y^2}{5x + 5y}$ *29* $\dfrac{a^2 - 7a - 18}{2a + 4}$ *30* $\dfrac{mn - n^2}{m^2 - n^2}$

31 $\dfrac{x^2-9}{(3x+9)(x^2-6x+9)}$ **32** $\dfrac{a}{b}\times\dfrac{a^2b-b^3}{a^3-ab^2}$ **33** $\dfrac{2u^2-3u-2}{u^2+u-6}$

34 $\dfrac{1}{x+1}+\dfrac{1}{x+2}$ **35** $\dfrac{2y-1}{2}-\dfrac{3y-2}{3}$ **36** $\dfrac{1}{x+1}-\dfrac{1}{x-1}$

37 $\dfrac{1}{v}-\dfrac{1}{u}$ **38** $\dfrac{a}{x}+\dfrac{b}{y}$ **39** $\dfrac{3}{x-5}-\dfrac{3}{x-2}$

40 A square of side $2r$ cm has semicircles drawn externally on each of two opposite sides. Find factorised expressions for:

  a the perimeter of the shape   *b*   the area of the shape.

Revision Exercise 1B

Factorise the expressions in questions *1–18*

1 $3x^2-6xy$ **2** x^2-x^6 **3** t^3+5t^2-24t

4 $(a-b)^2-100$ **5** $4(p+q)^2-9r^2$ **6** $x^2-(a-b)^2$

7 $x^2+2xy-24y^2$ **8** $4a^2-ab-3b^2$ **9** $3m^2+11mn+8m^2$

10 $42p^2-16p-10$ **11** $x^2-1-\dfrac{2}{x^2}$ **12** $\left(t+\dfrac{1}{T}\right)^2-\left(t+\dfrac{1}{T}\right)$

13 $x(x-1)+y(1-x)$ **14** $2x(3a+b)+3y(b+3a)$ **15** $1-16m^4$

16 $1-(p-q)^2$ **17** $x(x+4)-12$ **18** $c^2-(a+b)^2$

19   Calculate n given that $n=\sqrt{(5{\cdot}1^2-4{\cdot}5^2)}$.

20   Evaluate $\dfrac{\sqrt{(x^2-y^2)}}{z}$ exactly when $x=2{\cdot}5$, $y=2{\cdot}4$ and $z=3{\cdot}5$.

21   If $tu-tv=u^2+3uv-4v^2$, factorise both sides and hence express t in terms of u and v as simply as possible, assuming $u\neq v$.

22   Use the distributive law to simplify

$$ab\left(\dfrac{1}{a}+\dfrac{1}{b}\right)+bc\left(\dfrac{1}{b}+\dfrac{1}{c}\right)+ca\left(\dfrac{1}{c}+\dfrac{1}{a}\right),$$

giving your final result in factorised form.

23   Given that $x^2 - y^2 = 45$ and $x + y = 9$, find an equation for $x - y$ and hence find x and y.

24   If $x = p + 2q$ and $y = 2p + q$, show that:

a $x^2 - y^2 = 3(q^2 - p^2)$ b $x^2 + 2xy + y^2 = 9(p + q)^2$

Simplify the expressions in questions **25–35**:

25   $\dfrac{x^2 - y^2}{6x^2 - 5xy - y^2}$ 26 $\dfrac{a^2 + 2ab + b^2}{(a + b)^2}$ 27 $\dfrac{6ax + 6a}{3ax^2 + 3ax}$

28   $\dfrac{y - 2x}{2x^2 - 5xy + 2y^2} \times (2y - x)$ 29 $\dfrac{1}{x + 2} + \dfrac{2}{x^2 - 4}$

30   $\dfrac{1}{a^2 - ab} - \dfrac{b}{a - b}$ 31 $\dfrac{2}{a^2 - 9} - \dfrac{1}{a^2 + 3a}$

32   $\dfrac{1}{b(a - b)} + \dfrac{1}{a(b - a)}$ 33 $\dfrac{x + 3}{x - 1} - \dfrac{x - 1}{x + 3}$

34   $\dfrac{1}{x^2 - x - 2} - \dfrac{1}{x^2 + 4x + 3}$ 35 $\dfrac{1 + \dfrac{1}{a}}{a - \dfrac{1}{a}}$

36   A boiler consists of a cylinder, radius r metres and length h metres, with a hemisphere on each end. Find factorised expressions for its surface area and volume. Hence calculate the area and volume when $h = 15$, $r = 2.5$ and 3.14 is taken as an approximation for π. (Volume of cylinder $= \pi r^2 h$, volume of sphere $= \frac{4}{3}\pi r^3$, curved surface area of cylinder $= 2\pi rh$ and surface area of a sphere $= 4\pi r^2$.)

Revision Exercise on Chapter 2
The Language of Variation

Revision Exercise 2

1   Express each of the following as an algebraic equation:

a s varies as the square of t b p varies inversely as d

c $I \propto a^4$ d $P \propto \dfrac{1}{\sqrt{x}}$

2   Given that x varies directly as y and $x = 16$ when $y = 10$, find:
   a x when $y = 8$ b y when $x = 4$.

3   y varies inversely as x and $y = 8$ when $x = 6$. Find:

   a y when $x = 15$ b x when $y = 32$.

4   If $p \propto q^2$ and $p = 5$ when $q = 3$, find p when $q = 12$.

5   The quantities F, x and Q, v vary respectively in accordance with the
   information given in the following tables:

x	2	6	10		v	2	6	10
F	0·8	2·4	4·0		Q	30	10	6

   a Deduce how F varies with x, and how Q varies with v.
   b Find in each case the law of variation.

6   Given that Q varies directly as x and inversely as z, and $Q = 50$ when
   $x = 30$ and $z = 14$, find:

   a the formula connecting Q, x and z.
   b Q when $x = 36$ and $z = 42$.

7   The exposure E seconds required for a film varies directly as the
   square of the stop f used. It is found that $E = \frac{1}{100}$ when $f = 8$.

   a Obtain the equation connecting E and f.
   b Calculate E when $f = 16$.

8   Use the formula $y = \frac{2}{3}x$ to complete these ordered pairs:

   $$(6, \ldots), \ (-12, \ldots), \ (\ldots, 10), \ (\ldots, -2), \ (\ldots, 3·8)$$

9   The acceleration a m/s^2 of a car varies as the force F newtons
   provided by the engine. When $F = 2400$, $a = 2·5$. Calculate:

   a the acceleration provided by a force of 1600 newtons
   b the force necessary to give an acceleration of 4 m/s^2.

10   The electrical resistance, R ohms, of a wire is proportional to its
   length l m, and inversely proportional to the square of its diameter,
   d mm. When $l = 1$, and $d = 1$, $R = 0·5$. Calculate:

a the resistance of a wire of the same material of length 100 m and diameter 0·5 mm

b the diameter of a wire 1000 m long with a resistance of 125 ohms.

11 In a test on a metal-filament lamp, the voltages V gave the following resistances R:

V (volts)	60	75	90	100	110
R (ohms)	86	108	125	137	151

a Using a scale of 2 cm to 10 units on both axes and origin (60, 80), show that the graph of the relation is linear. Hence find an approximate formula in the form $R = aV + b$.

b Use this formula to calculate R when $V = 80$.

12 The velocity v m/s of a train, t seconds after the brakes are applied, is shown in the following table:

t	0	5	10	15	20	25	30
v	24	20	15	12	8	4·5	0

a Verify graphically that v and t are related by an approximate law of the form $v = at + b$.

b Determine a and b, and hence state the approximate law.

13 A current I produces a deflection $\theta°$ in an electrical measuring instrument such that I varies as $\tan \theta°$.

a When $I = 0·5$, $\theta = 18·5$. Show that an approximate formula for I is $I = 1·5 \tan \theta°$.

b Find I when $\theta = 37$.

c What is the deflection when $I = 2$?

14 R varies as the sum of two parts, one of which is constant and the other varies as V^2, so that $R = a + bV^2$. When $V = 2$, $R = 105$ and when $V = 3$, $R = 230$.

a Find a and b and hence state the formula.

b Calculate R when $V = 4$.

Revision Exercises on Chapter 3
Quadratic Equations and Inequations

(Assume that the variables are on the set of real numbers.)

Revision Exercise 3A

1   Write down the solution sets of the following:

a $(x-2)(x+5)=0$ b $(x+1)(x-1)=0$ c $(x-10)(x-\frac{1}{2})=0$

d $x(x-7)=0$ e $5x(x+10)=0$ f $(x-8)^2=0$

2   Use the method of factors to solve the following:

a $x^2+6x+8=0$ b $t^2-12t+20=0$ c $4y^2-49=0$

d $3x^2+5x-2=0$ e $40m-5m^2=0$ f $x(x+8)=65$

3   Find the constant c given that 4 is a root of the equation

$$2x^2+x+c=0.$$

Hence find the solution set of the equation.

4   The sides of a rectangle whose area is 126 cm^2 are $2x$ cm and $(x+2)$ cm in length. Form an equation in x and solve it to find x.

5   A parabola has equation $y=x^2-10x+16$.

a   Find the coordinates of the points where the parabola cuts the x-axis.

b   The point $(h, 16)$ lies on the parabola. Find two possible replacements for h.

c   The x-coordinate of the turning point is 5. Find its y-coordinate.

d   Use your information to sketch the parabola on graph paper.

6   Solve each of the following equations in x:

a $(x-5)^2=100$ b $(x+\frac{1}{2})^2=9$

c $(2x-7)^2=\frac{1}{4}$ d $(x-\frac{1}{4})^2=\frac{4}{25}$

7   The sum of the first n terms of the sequence 2, 6, 10, 14, ... is $2n^2$. How many terms of the sequence, beginning at 2, must be taken for the sum to be 338?

8 Solve each of the following equations, rounding off the roots to one decimal place:

 a $x^2 + 3x - 6 = 0$ *b* $2x^2 - 3x - 4 = 0$ *c* $5x^2 = 4 - 2x$

 d $t^2 + 0.4t - 1.3 = 0$ *e* $1 + 3x - 2x^2 = 0$ *f* $5x(x + 2) = 12$

9 The formula $W = VI - RI^2$ occurs in electrical theory. Find I when $W = 950,$ $V = 240$ and $R = 10$ so that $I < 10$.

10 Find the solution set of each of the following inequations:

 a $(x + 3)(x - 3) > 0$ *b* $(x - 5)(x - 8) \leqslant 0$

 c $x^2 - 7x + 12 \geqslant 0$ *d* $3 + 5x - 2x^2 > 0$

11 A sheet of metal is in the form of an isosceles triangle whose altitude is 2 cm shorter than the base of length x cm. If its area is less than 40 cm², show that $2 < x < 10$.

12 If p is a positive integer, find the smallest p for which $p(p + 1) > 1000$.

13a Solve the equation $x^2 - x - 6 = 0$.

 b Deduce the solution sets of: *(1)* $x^2 - x - 6 > 0$ *(2)* $x^2 - x - 6 < 0$.

14 The sides of a right-angled triangle are x cm, $(x + 7)$ cm and $(2x + 3)$ cm long, in ascending order. Find x and the area of the triangle.

15 The functions f and g from R to R are defined by the formulae $f(x) = x^2 + 2$ and $g(x) = 2x + 5$ respectively. Find x such that $f(x) = g(x)$.

Revision Exercise 3B

1 Find the solution set of each of the following equations:

 a $4x^2 + 12x - 27 = 0$ *b* $14 + x - 4x^2 = 0$

 c $9x^2 + 6x + 1 = 0$ *d* $2x^2 + x = 0$

2 Find the roots of the following equations:

 a $2(4t^2 - 11t) = 21$ *b* $(2n - 1)^2 - (3n - 1)(n - 3) = 5$

 c $(2p + 1)^2 = 2(p + 1)^2$ *d* $2x + \dfrac{8}{x} = 9 - \dfrac{2}{x}, \ x \neq 0$

3   Change the subject of the following formulae to the letter indicated:

a $eV = \frac{1}{2}mv^2$ to v *b* $E = \dfrac{q}{kr^2}$ to r

c $A = 4\pi r^2$ to r *d* $2\pi fL = \dfrac{1}{2\pi fC}$ to f

4   Find cos x from the equation $12\cos^2 x - \cos x - 6 = 0$.

5   The equation of a circle is $x^2 + y^2 - 5x + 2y - 24 = 0$. Find the coordinates of the points where the circle cuts:
a   the x-axis   *b*   the y-axis.

6   The function f is defined by $f(x) = 4(x-3)^2 + 5$, the domain of f being the set of real numbers.

a   If $f(x) = 21$, find x.   *b*   If the image of x is 5, find x.

7   Reduce each of the following equations to standard quadratic form, and hence find their solutions. Check the solutions in each case.

a $\dfrac{x}{x+6} = \dfrac{3}{x}$ *b* $\dfrac{10}{t} - \dfrac{10}{t+5} = 1$

c $\dfrac{1}{n} + \dfrac{1}{n+4} = \dfrac{1}{2}$ *d* $\dfrac{2x^2}{x+1} + 1 = \dfrac{2}{x+1}$

8   The formula $h = ut - \frac{1}{2}gt^2$ gives the height h of a body after time t when thrown vertically upwards with velocity u, and g is a constant. Calculate t when $g = 10$, $u = 25$ and $h = 20$. Explain the double answer.

9 *a*   Solve the equation $x^2 + 7x - 8 = 0$.
b   Deduce the solution sets of: (1) $x^2 + 7x - 8 > 0$ (2) $x^2 + 7x - 8 < 0$.

10   Solve the following equations in x, rounding off the roots to one decimal place.

a $6x^2 + 2x - 7 = 0$ *b* $10x^2 = 2x + 7$ *c* $5x^2 + 4x - 1 \cdot 3 = 0$

11   Use the quadratic formula to show that if $x^2 - 2ax + (a^2 - 1) = 0$, then $x = a + 1$ or $a - 1$.

12   Solve each of the following inequations and illustrate the solution set on the real number line.

a $x^2 + 6x - 40 < 0$ *b* $3 + 2x - x^2 \leqslant 0$ *c* $4x^2 - 4x + 1 > 0$

Cumulative Revision Section (Books 1-6)

Cumulative Revision Exercises

Cumulative Revision Exercise A

1 State which of the following are true and which are false:

a $(a-b)-c = a-(b-c)$ b If $A \subset B$, then $A \cup B = B$.

c $\dfrac{x}{4} - \dfrac{y}{2} = \dfrac{2x-y}{4}$ d If $2x = 3y$, then $\dfrac{x}{y} = \dfrac{3}{2}$.

e $(a-3b)^2 = a^2 + 9b^2$ f If $x = -1$, $x^3 + (x+2)^2 = 0$.

g If $x^2 + 2x - 15 = (x-3)(x-a)$, then $a = 5$.

2 $E = \{1, 2, 3, ..., 10\}$. List the following subsets of E:

a $\{x: 3 \leqslant x \leqslant 7\}$ b $\{x: x = 2p, p \in E\}$ c $\{x: x^2 \in E\}$

3 Simplify:

a $p^2 - q^2 + 3p^2 - 2q^2$ b $3(a-b) + 2(a+2b) - 5(a-b)$

c $5x^2 - 3x - 5 - (2x^2 - 3x - 1)$ d $(x+5)^2 - (x-1)^2$

4 Given $p = 3$, $q = -1$, $r = 0$, find the values of:

a $4pqr$ b $p^2 + q^2 + r^2$ c $2p^2 + 5q^2$ d $(2p + 5q)^2$

5 Show on Cartesian diagrams the solution set of each of the following, $x, y \in R$:

a $3x - 4y = 24$ b $y \geqslant 2$ c $y < x$ d $x + y \leqslant 6$

6 ⊕ means 'square the first number, then add the second'. * means 'add the two numbers, then square'. Simplify:

  a $1 \oplus 5$ *b* $5 \oplus 1$ *c* $p \oplus q$ *d* $1*5$ *e* $5*1$ *f* $p*q$

7 Find the solution sets of the following, $x \in R$:

  a $2x + 7 = 5$ *b* $3x - 8 > x - 2$ *c* $\frac{1}{2}x - \frac{1}{3}x = \frac{7}{4}$

  d $4(x+3) - 2(x+1) = 10$ *e* $\dfrac{x+3}{4} - \dfrac{x-2}{2} = 1$

8 Describe the shaded regions of the Venn diagrams in Figure 1 in terms of intersections and unions of the sets A, B and C.

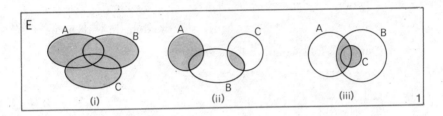

      (i)   (ii)   (iii)   1

9 Simplify:

  a $\dfrac{3 \pm 1 \cdot 8}{2}$ *b* $\dfrac{2 \pm 5 \cdot 6}{4}$ *c* $\dfrac{-3 \pm 1 \cdot 2}{10}$ *d* $\dfrac{-4 \pm 6 \cdot 7}{2}$

10 $E = \{a, b, c, ..., g\}$; $P = \{a, b, c, d, e\}$, $Q = \{c, d, e\}$, $R = \{d, e, f\}$. List the sets:

  a $P \cap Q \cap R$ *b* $P \cup Q \cup R$ *c* P' *d* $(Q \cap R)'$ *e* $Q' \cup R'$

11 Which of the following are true for the sets in question **10**?

  a $Q \subset P$ *b* $R \subset Q$ *c* $P \cap Q = R$ *d* $P' \subset Q'$

12 Simplify:

  a $7 - (-3) + 2$ *b* $7 \times (-3) \times 2$ *c* $-5 \times (-2) \times (-1)$

  d $(-3)^2 \times 0$

13 Solve the following systems of equations, $x, y \in R$:

  a $\left. \begin{array}{l} x - y = 3 \\ 2x + 3y = 11 \end{array} \right\}$ *b* $\left. \begin{array}{l} 5x - 2y = 12 \\ 7x - 3y = 17 \end{array} \right\}$ *c* $\left. \begin{array}{l} 2x + 5y + 2 = 0 \\ 3x - 4y + 3 = 0 \end{array} \right\}$

14 Which of the following are true and which are false?

a If y varies directly as x and inversely as z, then $y \propto x/z$.
b If the nth term of a sequence is $2n^2 - 1$, the sixth term is 143.
c $\{-2, -1, 0, 1\} = \{x: -2 < x < 1, x \in Z\}$
d The gradient of the line with equation $2x + y - 1 = 0$ is 2.
e If $A = \{(x, y): x + y = 0\}$, and $B = \{(x, y): x - y = 0\}$, where $x, y \in R$, then $A \cap B = \{(0, 0)\}$.

15 Factorise fully:

a $ab^2 - ac^2$ b $px^2 - py^2 + pz^2$ c $4x^2 - 1$ d $4x^2 - 9y^2$

e $x^2 - x - 12$ f $2x^2 - 5x + 2$ g $5p^2q - 5pq^2$ h $3a^2 - 10a + 7$

16 A function f is defined by $f(x) = 3x^2 + 1$, $x \in R$. Find:

a the image of -3 b the elements of the domain with image 13.

17 Expand the following products and squares:

a $(a + 5)(a - 1)$ b $(2x - 3)(3x - 2)$ c $(p + q)(p - q)$

d $(b + 4)^2$ e $(2c - 5)^2$ f $(5x - 4y)^2$

18 The length of a rectangle is $(x + 3)$ metres and the breadth is x metres. When the length is 2 metres more and the breadth is 1 metre less, the area of the rectangle is the same as before. Form an equation, and solve it to find x.

19 Figure 2 shows a region enclosed by the x- and y-axes and the lines with equations $y = 4$ and $x + y = 8$.

a Write down the coordinates of A, B, C and the point where CB produced would cut the y-axis.
b Give a set-builder definition of the region OABC.

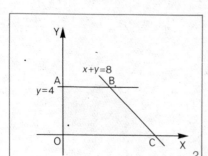

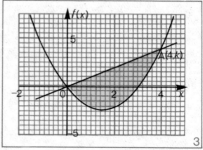

20 Part of the graph of the function $f:x \rightarrow x^2 - 3x$ is shown in Figure 3.

a What is the minimum turning value of the function?

b Write down the equation of the axis of symmetry of the graph.

c For what replacements for x is the value of f zero?

d Find the solution set of the inequation $f(x) < 0$.

e A straight line through O cuts the parabola at A$(4, k)$. Find k.

f Write down a set-builder definition of the shaded region.

21 Simplify: *a* $\dfrac{3x-9}{6}$ *b* $\dfrac{2}{2x+4}$ *c* $\dfrac{x^2-1}{3x+3}$ *d* $\dfrac{x^2-x-6}{x^2-4}$

22 $A = \{1, 3, 5, 7\}$. Describe each of these relations on A as a set of ordered pairs:

a is less than   *b* is equal to   *c* is greater than.

State the domain and range in each case, and illustrate each by an arrow diagram and by a Cartesian graph.

23 *a* y varies as x^2. Given $x = 3$ when $y = 36$, calculate:
(*1*) y when $x = 1$ (*2*) x when $y = 16$.

b p varies inversely as the square root of q. Given $p = 2$ when $q = 25$, calculate:
(*1*) p when $q = 4$ (*2*) q when $p = 5$.

24 A square has side $(x+4)$ cm long.

a Given that the perimeter is 50 cm, calculate x.

b Given that the area is 49 cm^2, calculate x.

25 $f(x) = 2x^2 + 3x - 2$, $x \in R$. Find the set of replacements for x for which:

a $f(x) = 0$ *b* $f(x) < 0$ *c* $f(x) > 0$.

26 Change the subject in each of these formulae to the variable stated:

a $C = \pi d$; d *b* $A = 2\pi rh$; h *c* $V = 4\pi r^2$; r

d $y = px + q$; x *e* $P = 2(l+b)$; l *f* $z = \dfrac{ky}{\sqrt{x}}$; x

27 Of 160 pupils, 84 chose sport, 66 art and 36 music in their 'choice' periods. None was able to take music and art, but 12 chose music and sport. Let x represent the number who chose art and sport, and show all this information in a Venn diagram. Hence find the number who chose art and sport.

28 Simplify:

a $\dfrac{1}{a}+\dfrac{1}{a-1}$ b $\dfrac{3}{u-3}-\dfrac{2}{u-2}$ c $\dfrac{x}{2x-1}-\dfrac{x}{2x+1}$

29 Solve the following quadratic equations, to 1 decimal place where necessary:

a $x^2+2x-8=0$ b $2x^2+7x+5=0$ c $2x^2+7x+4=0$

30 The volume of a cylinder varies directly as its length and as the square of its radius. Find the ratio of the volumes of two cylinders with lengths 25 cm and 15 cm, and radii 1·5 cm and 5 cm, respectively.

31 A function f is defined on R, except 3 and -3, by $f(x)=\dfrac{32}{x^2-9}$.

a Calculate the values of f at -1, 0 and 5.
b Find x such that $f(x)=2$.

32 Evaluate $5\sqrt{(p^2-q^2)}$ when $p=5\cdot6$ and $q=4\cdot4$, by first factorising p^2-q^2.

33 Find the solution set of each of the following for $x \in R$:

a $(x-2)^2-2(x-1)\leqslant x^2$ b $3(x-2)-(x-1)^2+1=0$.

34 The points $(2, -6)$ and $(3, 9)$ lie on the line with equation $y=mx+c$. By substituting the coordinates of the points in this equation, obtain two equations in m and c. Solve this system of equations to find m and c.

35a Draw the curve $y=x^2-2x-1$.
b Solve the equation $x^2-2x-1=0$ (1) using the graph (2) by another method.

36 The speed of an aircraft was noted after certain times as follows:

Time (T) in seconds	10	20	30	40	50	60
Speed (S) in m/s	32	50	69	94	110	132

Plot the corresponding points on squared paper, and draw the best-fitting straight line. Assuming that S and T are connected by a formula of the form $S=aT+b$, find a and b.

37 $A=\begin{pmatrix} 3 & -2 & 1 \\ 0 & -1 & 4 \end{pmatrix}$ and $B=\begin{pmatrix} 0 & -1 & 1 \\ -2 & 3 & 5 \end{pmatrix}$.

a Give the order of each matrix, and the element in the first row and second column of each.

b Simplify: (*1*) $A+B$ (*2*) $A-B$ (*3*) $3A+3B$

38 Find x and y, given that:

a $\begin{pmatrix} x & 3y \\ 1 & 2 \end{pmatrix} = \begin{pmatrix} -2 & -6 \\ 1 & 2 \end{pmatrix}$

b $\begin{pmatrix} 2x & 4 \\ -1 & 3y \end{pmatrix} + \begin{pmatrix} 8 & -4 \\ 1 & 0 \end{pmatrix} = \begin{pmatrix} 0 & 0 \\ 0 & 0 \end{pmatrix}$

39 Simplify the following matrix products:

a $(3 \ 5)\begin{pmatrix} 2 \\ -4 \end{pmatrix}$ *b* $(a \ b \ c)\begin{pmatrix} x \\ y \\ z \end{pmatrix}$ *c* $\begin{pmatrix} 1 & 4 \\ 0 & -1 \end{pmatrix}\begin{pmatrix} 2 & 0 \\ 1 & -1 \end{pmatrix}$

40 The point $(x, y) \rightarrow (x', y')$ under the mapping given by

$$\begin{pmatrix} x' \\ y' \end{pmatrix} = \begin{pmatrix} 2 & 0 \\ 0 & -1 \end{pmatrix}\begin{pmatrix} x \\ y \end{pmatrix}.$$

Find the images of the points $(1, 1)$, $(0, 3)$ and $(-2, -2)$ under the mapping.

41 Write down the inverse of each of the following matrices:

a $\begin{pmatrix} 3 & 2 \\ 4 & 3 \end{pmatrix}$ *b* $\begin{pmatrix} 1 & 0 \\ 0 & 1 \end{pmatrix}$ *c* $\begin{pmatrix} 2 & 0 \\ 0 & 2 \end{pmatrix}$ *d* $\begin{pmatrix} -2 & 3 \\ -4 & 5 \end{pmatrix}$

42 Use matrices to solve these systems of equations, $x, y \in R$:

a $\left. \begin{array}{l} x+y=3 \\ 2x-y=3 \end{array} \right\}$ *b* $\left. \begin{array}{l} 2x-5y=12 \\ 3x+2y=-1 \end{array} \right\}$ *c* $\left. \begin{array}{l} x-y-3=0 \\ 4x+3y-12=0 \end{array} \right\}$

Cumulative Revision Exercise B

1 Find the solution sets of the following, $x \in R$:

a $\frac{3}{4}x = 12$ *b* $-3x \leqslant -12$ *c* $2(3-x) > 5$

d $\frac{1}{2}x - \frac{1}{3}(x-4) > \frac{1}{6}(5x-7)$ *e* $(x-1)^2 - (x+1)^2 = 2(x+2)$

2 The nth term of a sequence is given by the formula $T = \frac{1}{2}n(n+1)$. Find: *a* the twentieth term *b* n when $T = 21$.

3   Expand the following:

a $(2x+4y)(6x-3y)$ b $(x^2-1)(x^2+1)$ c $(y^3+1)^2$

d $\left(x+\dfrac{1}{x}\right)^2$ e $(x+y)(x^2-xy+y^2)$ f $\left(x+1+\dfrac{1}{x}\right)\left(x-1+\dfrac{1}{x}\right)$

4   Write down the converse of each of the following. Then state whether the given statement is true or false, and whether the converse is true or false.

a $x=5 \implies x^2=25$ b $ab=0 \implies a=0$ and $b=0$

c $p>q \implies q<p$ d $\sin a°=0 \implies \cos a°=0$.

e If a triangle ABC is right-angled at A, then $a^2=b^2+c^2$.

5   Factorise fully:

a a^4-b^4 b px^3-px c $p^2-pq-20q^2$

d $2x^6-2x^2$ e $4a^2-10a-6$ f $(2a-3b)^2-1$

g $(2a-3b)^2-(3a-2b)^2$ h $a^2x^3+a^3x^2-a^2x^2$ i $\cos^2 A+2\cos A-3$

6   A rectangle is $2x$ metres long and x metres broad. Semicircles are drawn outwards on the longer sides. The perimeter of the resulting shape is P m, and its area is A m². Find factorised expressions for P and A, and hence calculate the perimeter and area when $x=5$ and $\pi=3{\cdot}14$.

7   Make x the subject of each of the following formulae:

a $xy=z(1-x)$ b $ax=\sqrt{(b^2-x^2)}$ c $d=\dfrac{b(x-a)}{c}$ d $\dfrac{1}{x}+\dfrac{1}{y}=\dfrac{1}{z}$

8 $f: x \to 1-x-x^2$. a Find the images under f of 2, 0 and -2.
b Find the elements of the domain for which the value of f is 1.

9 $x^2+ax+b=0$ when $x=-1$ and when $x=2$. Find a and b.

10   Simplify: a $\dfrac{x}{x+1}-\dfrac{2x}{2x-1}$ b $\dfrac{1}{a^2+4a+4}-\dfrac{1}{2a+4}$

11   The lines with equations $y=mx+c$ and $y=ax$ intersect at K. Find the coordinates of K in terms of a, m and c, assuming that $m \neq a$.

12 $C = \begin{pmatrix} -1 & 0 \\ 0 & 1 \end{pmatrix}$ and $D = \begin{pmatrix} 1 & 0 \\ 0 & -1 \end{pmatrix}.$

a Show that $CD = DC$.
b What geometrical transformations are represented by C and D?
c Interpret the result $CD = DC$ geometrically.

Geometry

Circle 1: Rotational Symmetry

1 Rotation—an introduction

Many natural and man-made objects in the world around us depend on rotation for their existence or their operation. Discuss the rotating parts and the centres of rotation, or axes of rotation, in the objects in Figure 1. You will need to think very carefully about the third row.

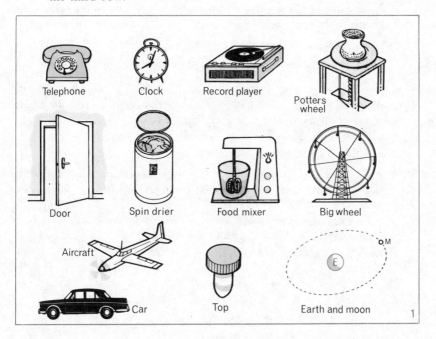

Telephone   Clock   Record player   Potters wheel

Door   Spin drier   Food mixer   Big wheel

Aircraft   Car   Top   Earth and moon

1

Exercise 1

Questions *1–3* refer to Figure 2 (i)–(iv), which show some simplified drawings of wheels with equally spaced spokes. O is the centre in each case.

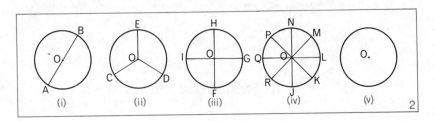

1   In parts (i) to (iv) make, or imagine, a tracing of each placed over the original drawing. As the tracing rotates about O, what happens to:

a   the tracings of the spokes
b   the tracing of the rim of the wheel?

2   What can you say about the distance of every point on the rim from the centre of the wheel?

3   Through which angles less than 360° can each tracing be rotated about O to make the spokes fit with those in the original drawing?

4   In Figure 2 (iii) the tracing fits the original after rotation through angles of 0°, 90°, 180° and 270°. So it fits (without turning over) in exactly four ways, and we say that it has *rotational symmetry of order 4*. Write down the order of rotational symmetry for Figures 2 (i), (ii) and (iv).

5   In Figure 2 (v), through how many angles less than 360° could you rotate the tracing about O so that it coincides with the original?

Since there is no limit to the number of ways in which a circle fits its outline under rotation about its centre, the circle is said to have rotational symmetry of infinite order.

2 Rotation—a transformation of the plane

We have already studied the transformations of points of a plane as illustrated in Figure 3.
(i)   Reflection in an axis of symmetry (Book 3)
(ii) Translation through a given distance in a given direction (Book 4)

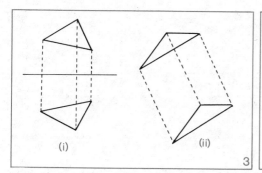

(i)   (ii)   3

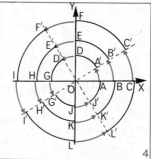

4

We can think of rotation as a transformation of all points of the plane as they move through arcs of circles about the centre of rotation.

In Figure 4, OA→OA', OB→OB', OC→OC', etc.,

and   A→A', B→B', C→C', etc.

When we describe a rotation we must always think of three things: the centre of rotation, the magnitude and the direction of the rotation.

As in Book 5, Trigonometry, we think of anticlockwise rotations as positive, and clockwise rotations as negative, particularly in coordinate diagrams.

Exercise 2

1   Copy Figure 4, showing three concentric circles with centre O, and the points A,B,C,...,L. Mark the images of these points after an anticlockwise (positive) rotation about O of 30°, and then after a further 30°.

2   In Figure 5, a rotation of 90° about O has mapped △OAB to
    △OA′B′.

a   How would you describe the direction of the rotation?
b   Name equal lines and angles in the figure.
c   What can you say about triangles OAB and OA′B′?
d   Through what angle has (1) OA (2) OB (3) AB turned?
e   Which point remains fixed (invariant) under the rotation?
f   Copy the diagram and sketch the image of △OAB under negative
    (clockwise) rotations of 90°, 180° and 270°.

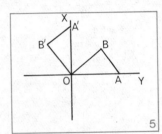

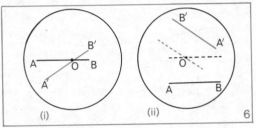

3   Figure 6 (i) shows a plan of a record-player turntable, with a chalk
    line AB marked on it and its image A′B′ after an anticlockwise
    rotation about O of 40°.

a   Name a line equal to OA and one equal to OB.
b   State the sizes of angles AOA′ and BOB′.
c   What is the size of the acute angle between AB and its image?

4   Repeat question 3 for Figure 6 (ii) where the turntable rotates anti-
    clockwise through 140° about O. Explain carefully your answer to
    part c.

If A→A′ and B→B′ under a rotation
of x° about O (see Figure 7), then
(i)   AB→A′B′
(ii) triangles OAB and OA′B′ are
      congruent
(iii) the angle between any line and its
      image is x°
(iv) O is an invariant point (i.e. O
      does not change its position).

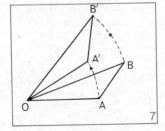

5   △ABC has vertices (4, 0), B (6, 0) and C (5, 4). On the same diagram
on squared paper draw △ABC and its images after positive rotations about O of: *a* 90°   *b* 180°   *c* 270°.
    Produce the sides as far as necessary to show the angles through
which each side of triangle ABC has turned.

6   P is the point (5, 5). Write down the coordinates of its images under
rotations about O through:

    a 90°   *b* −90°   *c* 180°   *d* 45°   *e* −45°.

7   ABCD is a square, named in the anticlockwise direction, whose
diagonals intersect at O. Write down the image of B under anticlockwise rotations of 90° about: *a* O   *b* A.

8   PQRS is a parallelogram whose diagonals intersect at M. Write
down the images of P, Q and M under rotations of 180° about M.
    Why can the direction of rotation be omitted in this case?

9   In Figure 8 a rotation about its centre O maps equilateral triangle
PQR to triangle P′Q′R′ so that R′Q′∥QR. Sketch the diagram, and
mark the centre of rotation. Through what angle has each of the
following turned? *a* PQ   *b* RQ   *c* △PQR

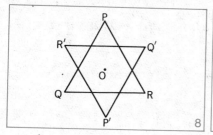

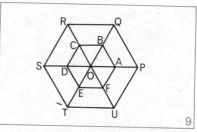

10   Figure 9 is the view from above of the wires of a rotating clothes-
drying frame. The angles at O are all equal. Under a rotation about
O, A→C. To what do the following map?

    a P   *b* C   *c* T   *d* OF   *e* RS

11a   In Figure 9, through which angles less than 360° can this figure be
rotated so as to coincide with itself?

    b What is its order of rotational symmetry?

12a Suppose you have a rectangular card which looks rather like a square. How could you test whether it is a square, after drawing its outline on a sheet of paper?

b Suppose you have a plate which is meant to be circular but you suspect it is not. How could you test after drawing its outline on a sheet of paper?

c In Figure 10 one of the arcs is intended to show part of a circle. Make a tracing of each arc and use these to find which is the circular arc.

Then do what is necessary to obtain on your tracing paper a drawing of the complete circle. (Use nothing but Figure 10, the tracing paper and a pencil.)

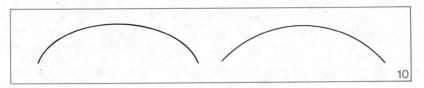

10

3 Chords, arcs and sectors of a circle

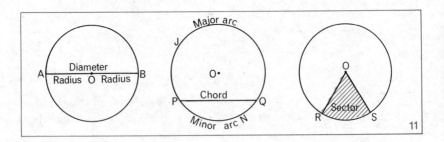

11

Exercise 3

1 In Figure 11, O is the centre of each circle, and different parts of the circle are labelled. Study the figure and use the given letters to list the names of:

a a diameter   *b* four radii   *c* a chord

d a major arc   *e* a minor arc   *f* a sector.

2 *a* Is every diameter a chord? Is every chord a diameter?

b Which of the following is true?

(*1*) The set of chords is a subset of the set of diameters.

(*2*) The set of diameters is a subset of the set of chords.

3   Draw three circles with centres A, B and C.

a In the first circle draw a chord PQ equal in length to the radius. Describe the triangle APQ.

b In the second draw a chord RS shorter than the radius. Describe the triangle BRS.

c In the third draw a chord TU longer than the radius. Describe the triangle CTU. Is it obtuse-angled? Must it be obtuse-angled?

4 *a* What does 'diametrically opposite' mean?

b Supply the missing words in the following sentence.

'A ...... of a circle is a straight line segment which joins two points on the ...... and passes through the .... Its length is double that of the ......,'

5 *a* Draw a circle, centre O, and mark points M and N on the circumference, not diametrically opposite each other.

b Label the major arc and the minor arc.

c Join OM and ON. Shade two sectors in the circle.

---

4 Relations between arcs, sectors and angles

A *sector* of a circle is an area bounded by two radii and an arc of the circle (see the shaded area OAB in Figure 12). Two radii divide the circle into two sectors, called the *minor* (smaller) sector and the *major* (greater) sector.

Exercise 4

1   In Figure 12, a tracing of sector OAB has been rotated about O to the position OA'B'. What can you say about:

a the sizes of angles AOB and A′OB′
b the lengths of arcs AB and A′B′
c the areas of sectors OAB and OA′B′?

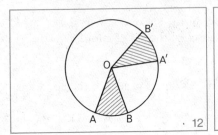

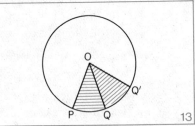

2   In Figure 13, a tracing of sector OPQ has been rotated about O to the position OQQ′. What can you say about:

a the sizes of angles POQ and QOQ′
b the lengths of arcs PQ and QQ′
c the areas of sectors OPQ and OQQ′
d the ratios (*1*) arc PQ: arc PQ′ (*2*) angle POQ: angle POQ′
    (*3*) sector OPQ: sector OPQ′?

3 *a* Draw a circle with centre O. Mark points A and B on the circumference so that the size of angle AOB is not more than 30°.
  b Mark points C, D, E and F on the circumference so that
       ∠AOB = ∠BOC = ∠COD = ∠DOE = ∠EOF.
  c Given that ∠AOB = x°, write down the sizes of angles BOC, COE and AOF.
  d Given that arc AB is y cm long, write down the lengths of arcs BC, CE and AF.
  e Given that the area of sector AOB is z cm², write down the areas of sectors BOC, COE and AOF.
  f Write down the numerical values of the ratios:

      (*1*) ∠BOC: ∠COE: ∠AOF   (*2*) arc BC: arc CE: arc AF

      (*3*) sector BOC: sector COE: sector AOF.

   The angle *subtended* by an arc at the centre of a circle is the angle at the centre facing the arc, e.g. in Figure 14, arc CD subtends the angle COD.

In a circle, the length of an arc and the area of a sector are proportional to the size of the angle subtended at the centre of the circle; for example in Figure 14,

$$\frac{\text{arc AD}}{\text{arc DC}} = \frac{\angle \text{AOD}}{\angle \text{DOC}} = \frac{\text{sector AOD}}{\text{sector DOC}},$$

4   In Figure 14, where O is the centre, find the numerical values of the ratios:

a $\angle$ AOD: $\angle$ DOC   b $\angle$ AOD: $\angle$ AOC   c $\angle$ DOC: $\angle$ COB

d arc AD: arc DC   e arc AD: arc AC   f arc AD: circumference

g sector AOD: sector DOC   h sector DOB: circle.

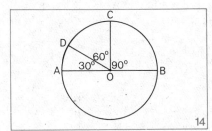

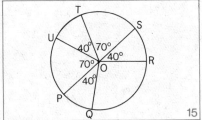

5   In Figure 14, given that the length of arc AD is 4 cm, calculate the lengths of arc DC, arc CB and the circumference of the circle.

6   In Figure 14, given that the area of sector DOC is 12 cm², calculate the area of the circle.

7   In Figure 15, where O is the centre:

a name three equal arcs
b calculate the length of the arc PQ, if the circumference of the circle is 54 m
c calculate the area of the sector QOR, if the area of the circle is 36 cm².

8   The tip of the minute hand of a clock moves round the circumference of a circle on which the 60 minute intervals are marked. Which of the following statements are true and which are false?

a The arc joining '5 past' to '20 past' has the same length as the arc joining 'half past' to 'quarter to'.

b   The hand turns through an angle of 3° every minute.

c   If the arc joining '10 past' to 'quarter past' measures 4 cm, the arc joining '10 past' to '25 past' measures 12 cm.

d   The area of the sector swept out by the hand in 1 hour is four times the area swept out between 'quarter past' and 'half past'.

e   As the hand turns through 72°, the tip passes over $\frac{1}{8}$ of the circumference.

9   Figure 16 shows a circle, centre O, with diameter AOC and radius OB such that $\angle COB = 45°$.

a   Calculate the ratio $\angle BOC : \angle AOB$.

b   Calculate what fraction arc CB is of arc BA. If arc BA is 12 cm long, what is the length of arc CB?

c   What fraction is sector BOC of the semicircle ABC? If the area of sector COB is 54 cm², calculate the area of the semicircle.

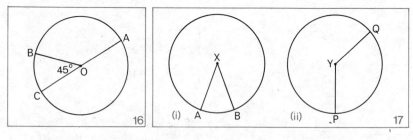

10   Figure 17 shows two circles with centres X and Y and equal radii.

a   If $\angle AXB = 20°$, $\angle PYQ = 120°$ and arc AB is 3 cm long, calculate the length of arc PQ.

b   If arc AB = 2 cm, arc PQ = 10 cm and $\angle PYQ = 140°$, calculate $\angle AXB$.

c   If $\angle AXB = x°$, $\angle PYQ = y°$, area of sector AXB = a cm² and area of sector PYQ = p cm², write down two equal ratios.

5 Equal chords in a circle

In Figure 18, the arc AB has been rotated about O into the position A'B'. Lines OP and OP' are perpendicular to AB and A'B' respectively. We then have

$$\angle \text{AOB} \rightarrow \angle \text{A}'\text{OB}'$$
$$\triangle \text{AOB} \rightarrow \triangle \text{A}'\text{OB}'$$
chord AB→chord A'B', hence AB = A'B'
OP→OP', hence OP = OP'.

We obtain the following results for every circle.

Arcs are equal
⇔ angles at the centre are equal
⇔ chords are equal
⇔ chords are equidistant from the centre.

When we say that equal chords cut off equal arcs, we must be careful to choose the right pair of arcs, both minor or both major.

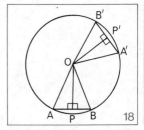
18

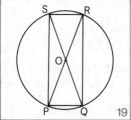
19

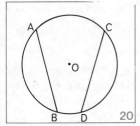
20

Exercise 5

1   Copy Figure 19, in which POR and QOS are diameters.

a Name the angle at the centre subtended by the chord PQ (i.e. the angle at the centre facing PQ).

b Name the angle at the centre subtended by the chord SR.

    c Why are these angles equal? What can you say about the chords PQ and SR?

    d Repeat *a*, *b* and *c* for chords PS and QR.

2   In Figure 20, chord AB = chord CD. Copy the figure. Draw and name two angles at the centre of the circle subtended by the chords. Why are the angles equal?

3   In your diagram for question *2* draw two lines which show the distance from O to the chords. Why are these lines equal?

4   In Figure 21, O is the centre of the circle and XY = XZ. Name:

    a a line equal to ON   *b* an angle equal to ∠ XOY
    c an arc equal to arc XY   *d* a sector equal to sector OZX.

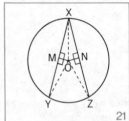
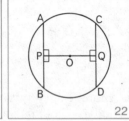
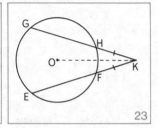

5   In Figure 22, AB and CD are equal chords in a circle centre O. POQ is a straight line. OP = 3 cm and CD = 5 cm.

    a Write down the lengths of PQ and AB.

    b Why is AB parallel to CD?

6   In Figure 23, EF and GH are equidistant from the centre O. Given that GHK and EFK are straight lines, with KF = KH, explain why KE = KG.

7   In Figure 24, SP = NP and MP = RP. Explain why chords MN and RS must be equidistant from O, the centre of the circle.

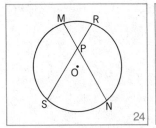

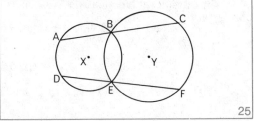

24

25

Exercise 5B

1 In Figure 25, ABC and DEF are straight lines, and X and Y are the centres of the two intersecting circles. Given that AB and DE are equidistant from X and that AC = DF, prove that BC and EF are equidistant from Y.

2 In a circle ABC with centre O, angles AOB and BOC are equal. Which of the following are true and which are false?

  a Arc ABC = twice arc AB.   b Chord AC > twice chord AB.

  c Chord AC = twice chord AB.   d Chord AC < twice chord AB.

3 a If the tip of the minute hand of a clock traces out $\frac{3}{8}$ of the circumference of a circle, through what angle does the hand turn?

  b If the minute hand turns through 48°, what fraction of the circumference does the tip of the hand trace out?

  c If the tip of the *minute* hand traces out $\frac{3}{4}$ of the circumference of a circle, through what angle does the *hour* hand turn?

4 In a circle with centre O, AB is a chord equal to the radius of the circle. Explain why the arc AB is $\frac{1}{6}$ of the circumference of the circle in length.

5 A circle has centre O and 10 equally spaced radii.

  a Under an anticlockwise rotation about O, through which angles between 0° and 360° will the figure fit into its own space?

  b Under a negative (clockwise) rotation about O, through which angles between −90° and −270° will it fit into its own space?

  c If the ends of the radii are joined to form a polygon, calculate the size of one of its angles.

6 An equilateral triangle ABC has centre O.

a Under which of the following rotations about O does it map onto itself?

0°, 30°, 90°, 120°, 180°, 240°, 300°

b If A→C under one of these rotations, complete the following.

B→..., OC→..., △AOC→...

c Sketch the triangle ABC and its image A'B'C' after an anticlockwise rotation of 60° about O. What can you say about the sides and angles of figure AA'BB'CC'?

6 Regular polygons

Exercise 6

1 Figure 26 shows four circles with centres O and equally spaced radii. The ends of the radii are joined to form shapes bounded by straight lines.

a By considering rotations about O, explain why these shapes have their sides equal and their angles equal.
b By making tracings and rotating them, verify that the sides and angles are equal.
c State the order of rotational symmetry of each shape.
d Name the shapes in Figures 26 (i) and (ii).

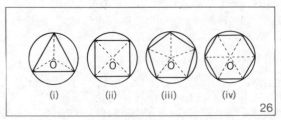

(i)   (ii)   (iii)   (iv)

26

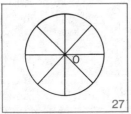

27

A figure with more than four sides is called a *polygon*. When all its sides are equal and all its angles are equal, it is called a *regular polygon.* Polygons have special names according to the number of sides, e.g. *pentagon* (5 sides), *hexagon* (6 sides), *octagon* (8 sides), *decagon* (10 sides).

Figures 26 (iii) and (iv) show a regular pentagon and a regular hexagon respectively.

2   In Figure 27, the radii are equally spaced round the circle. Explain how to obtain a regular octagon in the figure.

3 *a*   Calculate the size of the equal angles at the centre of each shape in Figure 26.

  b   Use your compasses, ruler and protractor to construct the shapes.

4   In Figure 26 (iii) each angle at the centre is 72°. Hence in each isosceles triangle in the figure the equal angles are $\frac{1}{2}(180-72)° = 54°$. Copy and complete the following table.

Polygon	Angle at centre	Base angles of isosceles △	Angle of polygon	Order of rotational symmetry
Pentagon	72°	54°	108°	5
Hexagon	60°	60°		
Dodecagon (12 sides)	30°			
100-sided polygon				
n-sided polygon				

5   Construct a regular pentagon with each side 4 cm long. (This time you cannot start with a circle. Think about one of the five isosceles triangles, calculate its angles and construct it first, then draw the circle.)

Summary

1 A circle has *rotational symmetry* of infinite order about its centre.

2 If under a *rotation of the plane* through $x°$ about O, A→A′ and B→B′, then

   a AB→A′B′
   b triangles OAB and OA′B′ are congruent
   c the angle between any line and its image is $x°$
   d O is an invariant point

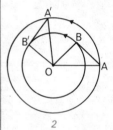

2

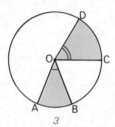

3

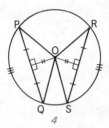

4

3 In a circle the *length of an arc* and the *area of a sector* are proportional to the *size of the angle subtended by the arc at the centre of the circle.*

$$\frac{\text{arc AB}}{\text{arc CD}} = \frac{\angle AOB}{\angle COD} = \frac{\text{sector AOB}}{\text{sector COD}}$$

4 *Equal arcs, chords and angles at the centre of a circle*
   Arcs are equal
   ⇔ angles at the centre are equal
   ⇔ chords are equal
   ⇔ chords are equidistant from the centre.

5 *Regular polygons* have all their sides and angles equal. They have rotational symmetry of certain orders about their centres.

Circle 2: Bilateral Symmetry

1 Bilateral symmetry

The 'Big Wheel' of the fairground (Figure 1) is not really circular, though it is called a wheel; it rotates about a centre and it possesses rotational symmetry. The regular polygons we met in Chapter 1, Section 6, might be regarded as simplified 'Big Wheels'. They have *rotational symmetry*, but they also have *line symmetry* or *bilateral symmetry*.

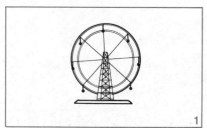

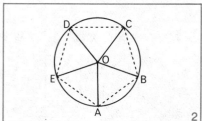

Exercise 1

1   A tracing of the 'spoked wheel' in Figure 2, where the angles at the centre O are all equal, can be picked up, *turned over*, and fitted on the original figure.

   a   In how many ways can this be done?
   b   How many axes of symmetry has the regular pentagon ABCDE?
   c   How many of these are also axes of symmetry of the circle?

2   Repeat question *1* for:

   a   a square
   b   a regular hexagon inscribed in a circle.

113

In every circle we can draw a regular polygon with any number of sides that we choose, and the resulting figure will possess all the rotational and bilateral symmetries of the polygon.

We now look more closely at the bilateral symmetry of a circle. In Figure 3 (i), O is the centre of the circle, AB is a chord and XY is the diameter perpendicular to AB.

Since OA = OB (radii of the same circle), △OAB is isosceles.

Since XY is perpendicular to AB, XY is an axis of symmetry of △AOB.

So XY bisects AB at right angles.

In the same way, XY bisects at right angles all chords parallel to AB, i.e. XY is an axis of symmetry of the circle; the part of the circle on one side of XY is the reflection in XY of the part on the other side.

So every diameter in a circle is an axis of bilateral symmetry.

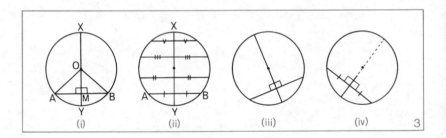

(i)   (ii)   (iii)   (iv)   3

Since the above results are true for every position of AB, it follows that:

(i)   every diameter is an axis of bilateral symmetry
(ii) a diameter (radius) perpendicular to a chord bisects it (Figure 3 (iii))
(iii) a diameter (radius) which bisects a chord is perpendicular to it (Figure 3 (iii))
(iv) the perpendicular bisector of a chord passes through the centre (Figure 3 (iv)).

Exercise 2

1 a Draw a circle with an axis of symmetry PQ.
  b Draw a chord CD so that PQ is still an axis of symmetry of the figure. What can you say about CD?

2 a Draw a circle with a chord MN.
  b Draw a line ST which is an axis of symmetry of the figure. What can you say about ST and MN?

3 In Figure 4, O is the centre of the circle and XY is perpendicular to PQ.

  a Name the axis of bilateral symmetry of the figure.
  b If ∠ POR = 55°, copy the diagram and fill in the sizes of all the angles.
  c If PQ = 10 cm, write down the lengths of PR and RQ.
  d Name two congruent triangles in the figure.
  e Name two pairs of equal arcs in the figure.

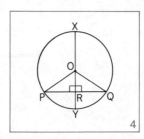 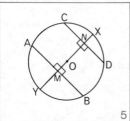 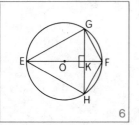

4 In Figure 5, chords AB and CD are both perpendicular to XY. Name: *a* the axis of symmetry *b* two pairs of equal lines *c* three pairs of equal arcs.

5 In Figure 6, chord GH is perpendicular to diameter EF. Copy the figure and mark equal lines, angles and arcs. What kind of figure is EHFG?

6 ST is a diameter of a circle and a line XTY is drawn perpendicular to ST. A chord PQ is drawn parallel to XTY to cut ST at K.

  a Name the axis of symmetry.
  b Name two equal chords you could draw in the figure.

c   Name two equal arcs in the figure.

d   Name four equal angles.

7 *a*   Draw a circle, centre O, and two parallel chords AB and CD. Draw the axis of symmetry of the figure.

b   Name a chord equal to AC, an arc equal to arc BD and an angle equal to ∠ACD.

c   If AD and BC are joined, name three angles each equal to ∠ADC.

Exercise 2B

1   Draw a circle, centre O, and mark a point A inside the circle. Show how to draw accurately a chord which will be bisected at A.

2   AB and CD are two chords in a circle with OM and ON their perpendicular distances from the centre. Use Pythagoras' theorem in triangles AOM and CON to show that $AM^2 + OM^2 = CN^2 + ON^2$.

What can you deduce about the chords AB and CD if:

a $OM = ON$ *b* $OM > ON$ *c* $OM < ON$?

3   In the same figure as in question *2*, what can you deduce about OM and ON if:

a $AB = CD$ *b* $AB > CD$ *c* $AB < CD$?

4   Use the results of questions *2* and *3* to complete the following statements.

a   Of two unequal chords in a circle, the greater is ....

b   Of two unequal chords in a circle, that which is nearer the centre is ....

5   In a circle, centre O, AL and BM are equal arcs where L and M are on the same side of AB. XY is the diameter perpendicular to AB. Use XY as an axis of symmetry to prove that LM is parallel to AB.

2 Finding the centre of a circle; the circumcircle of a triangle

Exercise 3

1   On tracing paper draw a circle without its centre (e.g. draw round a plate or a circular tin). Find the positions of two diameters by folding, and hence mark the centre of the circle.

2   Trace again the circular arc of Figure 10, page 102. Find, by folding, the centre of the circle of which it is a part; complete the circle using compasses.

3 *a*   A circle has to be drawn through two points A and B. On which line must the centre of the circle lie? (Remember that AB is a chord of the circle.)

b   How many circles can be drawn through A and B? What is the locus of their centres?

c   Is there a smallest circle that can be drawn through A and B? Is there a largest?

d   For a given radius of length *r* units discuss how many circles can be drawn through A and B.

4   Mark three points A, B and C which are *not* collinear (i.e. not in a straight line) on a sheet of paper.

a   Where are the centres of all possible circles through A and B?

b   Where are the centres of all possible circles through B and C?

c   If it is possible, draw a circle, or circles, through A, B and C.

d   If three points are not collinear, is it always possible to draw a circle through them? How many such circles can be drawn for the same three points?

5   Explain why it is not possible to draw a circle through three collinear points.

6 *a*   Draw an acute-angled triangle ABC. Find the centre O of the circle which passes through A, B and C and draw the circle.

b   Repeat for an obtuse-angled triangle ABC.

c   Repeat for a right-angled triangle ABC.

d   What can you say about the position of the centre in each of the three cases?

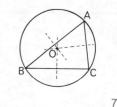

Note. The circle through the vertices of triangle ABC is called the circumcircle of triangle ABC; its centre O is called the circumcentre of triangle ABC (see Figure 7).

7

7   In triangle ABC, the perpendicular bisectors of AB and BC meet at O. Why must O lie on the perpendicular bisector of AC? What can you say about the perpendicular bisectors of the three sides of the triangle?

8   Mark any four points A, B, C and D on a sheet of paper so that no three of them are collinear.

a   Draw the circle through A, B and C.
b   Repeat for the points B, C, D; C, D, A; D, A, B.
c   Is it always possible to draw a circle through A, B, C and D?
d   Would it ever be possible to draw a circle through A, B, C and D?

Note. When one circle can be drawn through four (or more) points they are said to be *concyclic*. These points have special properties which will be discussed in Book 7.

9   Discuss the possibility of drawing a circle through the vertices of:
a   a parallelogram   *b*   a rectangle   *c*   a rhombus
d   a square   *e*   a kite.

3 Calculations using Pythagoras' theorem

Example. Figure 8 shows a circle with centre O and radius 13 cm. Chord AB is 24 cm long and is perpendicular to diameter XY. Calculate:

a   the length of OM   *b*   the length of MX
c   the area of △AXB

a XY is the axis of symmetry $\Rightarrow AM = MB = \frac{1}{2}AB = 12$ cm.
By Pythagoras' theorem in $\triangle OAM$,

$$OM^2 = OA^2 - AM^2$$
$$= 13^2 - 12^2 \text{ cm}^2.$$
$$= 25 \text{ cm}^2.$$

Hence $OM = 5$ cm.

b $MX = MO + OX = 5 + 13$ cm $= 18$ cm.

c Area of $\triangle AXB = \frac{1}{2}AB.MX = \frac{1}{2} \times 24 \times 18$ cm$^2 = 216$ cm^2.

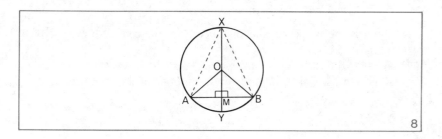

8

Exercise 4

1 In Figure 8, if $OA = 10$ cm and $AB = 16$ cm, calculate:

a the length of OM   *b* the length of MX   *c* the area of $\triangle AXB$.

2 Repeat question *1* if in Figure 8 $OA = 25$ m and $AB = 48$ m.

3 In Figure 9, $AO = 17$ cm and $OM = ON = 8$ cm. Calculate the lengths of: *a* AM *b* AB *c* CD.

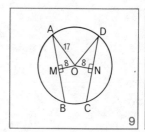

9

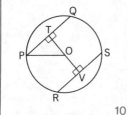

10

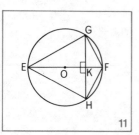

11

4   In Figure 10, PQ = RS = 24 m and OP = 15 m. TOV is a straight line. Calculate the lengths of: *a* PT *b* OT *c* TV.

5   In Figure 11, EF is perpendicular to GH. The radius of the circle is 2 m and GK is 1·6 m long. Calculate the lengths of OK and KF.

6   A chord AB, 10 cm long, is 12 cm from the centre O of a circle. Calculate the length of the radius of the circle and the size of angle OAB.

7   In a circle, centre O, the radius is 10 cm. If chords AB and CD are respectively 3 cm and 4 cm from O, calculate the lengths of AB and CD to three significant figures.

8   A circle has radius 3·4 cm.

 a What is the perpendicular distance from the centre to every chord 6 cm long?
 b What is the locus of the midpoints of such chords?
 c Illustrate by a diagram showing a number of these chords.

Exercise 4B

1   PQ and RS are two parallel chords in a circle with centre O and radius 10 cm. If PQ = 16 cm and RS = 12 cm, calculate the distance between PQ and RS. Can you find more than one answer?

2   Figure 12 illustrates the penalty box on a football pitch. The penalty spot O is 11 m from the goal line, and the line AB is 16 m from the goal line. A circular arc, with centre O and radius 9 m is marked. Calculate the length of the line AB to the nearest metre.

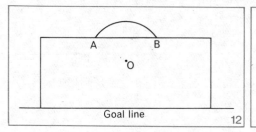

12

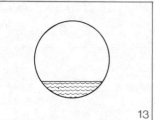

13

3   Figure 13 shows a section of a horizontal cylindrical pipe of *diameter* 1 m. Water is lying in the pipe and is 20 cm deep at the middle of the section. Calculate the width of the water surface.

4   A water pipe (as in question *3*) has a *diameter* of 80 cm. The width of the water surface is 48 cm. Show that the depth of water could be 8 cm. Is it possible to find another answer?

5 *a* For a circle of radius 10 cm, copy and complete the following table for chords, working to three significant figures.

Distance from centre (cm)	0	1	2	3	4	5	6	7	8	9	10
Length of chord (cm)		19·9	19·6	19·1		17·3					

b   Draw a graph of length of chord against distance from centre.
c   From your graph estimate the length of a chord 6·5 cm distant from the centre, and the distance from the centre of a chord 15 cm long.
d   Check the accuracy of your answers to *c* by calculation.

---

4 The equation of a circle

Exercise 5

1   Figure 14 shows a circle with centre O and radius 6 units.

a   Why is   (*1*) $a^2 + b^2 = 36$ (*2*) $c^2 + d^2 = 36$?
b   Write down similar equations for the points S and T.

2   Figure 15 shows a point P (x, y) on a circle with centre O and radius 5 units. Write down an equation in x and y.

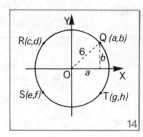

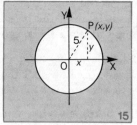

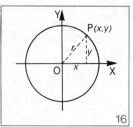

3 a In Figure 15, where will we find all points for which $x^2 + y^2 = 25$?
b Where will we find all points for which $x^2 + y^2 > 25$?
c Where will we find all points for which $x^2 + y^2 < 25$?

4 Figure 16 shows a point P (x, y) on a circle with centre O and radius r. Write down an equation in x, y and r.

For a point P(x, y), the equation $x^2 + y^2 = r^2$ is true if P lies on the circumference of the circle centre O, radius r; the equation is not true if P is anywhere else. We say that $x^2 + y^2 = r^2$ is the *equation of the circle*, centre O, radius r.

$\{(x, y): x^2 + y^2 = r^2\}$ is the set of points on the circumference of the circle.

$\{(x, y): x^2 + y^2 < r^2\}$ is the set of points inside the circle.

$\{(x, y): x^2 + y^2 > r^2\}$ is the set of points outside the circle.

Exercise 6

1 Write down the equations of the circles with centre O and radii 2, 5, 6, 10 and 12 units.

2 Write down the lengths of the radii of circles with equations:

a $x^2 + y^2 = 4$ *b* $x^2 + y^2 = 49$ *c* $x^2 + y^2 = 1$ *d* $x^2 + y^2 = 400$.

3 Write down an equation or inequation for each of the following statements.

a The point P(x, y) is 3 units from the origin.
b P(x, y) is more than 10 units from O.
c P(x, y) lies within the circle, centre O, radius 8 units.

4   Write down the centres and radii of circles with equations:

   a $x^2 + y^2 = 16$ *b* $x^2 + y^2 = 100$
   c $x^2 + y^2 = a^2$ *d* $x^2 + y^2 = 0.25$.

5   Draw a circle with centre O and radius 3 units. Which of the following is true for every point on its circumference?

   a $x + y = 3$ *b* $x + y = 9$ *c* $x^2 + y^2 = 3$ *d* $x^2 + y^2 = 9$

6   In your diagram for question *5* where will you find all the points for which:

   a $x^2 + y^2 = 9$ *b* $x^2 + y^2 > 9$ *c* $x^2 + y^2 < 9$?

7   Write down an equation or an inequation for each of the following.

   a The locus of points P(x, y) which are 9 units from the origin.
   b The locus of points P(x, y) which are more than 8 units from O.
   c The locus of points P(x, y) which lie inside the circle with centre O and radius 6 units.

8   Indicate on squared paper the locus of points given by:

   a $\{(x, y): x^2 + y^2 = 25\} \cap \{(x, y): x = 3\}$

   b $\{(x, y): x^2 + y^2 = 25\} \cup \{(x, y): x = 3\}$

   c $\{(x, y): x^2 + y^2 \geqslant 9\}$ *d* $\{(x, y): x^2 + y^2 \leqslant 25\}$

   e $\{(x, y): x^2 + y^2 \geqslant 9\} \cap \{(x, y): x^2 + y^2 \leqslant 25\}$

   f $\{(x, y): x^2 + y^2 \leqslant 16\} \cap \{(x, y): y \geqslant x\}$

9   What can you say about the set of points such that $x^2 + y^2 = -10$?

10a   Work out the distance of the point A(3, 4) from the origin. What is the equation of the circle, centre O, which passes through A?

   b B(3, p) is another point on this circle. What is the value of p?
      C(q, 4) is another point on this circle. What is the value of q?

   c Using the bilateral symmetry of the circle, write down the coordinates of another point on the circumference.

   d If (4, r) is a point on the circumference, write down two possible replacements for r. Repeat for s, t and u if the points (−4, s), (t, 3) and (u, −3) lie on the circumference.

   e Plot all the points obtained so far. Insert in the diagram all the axes of symmetry of this set of points. Does the set of points possess rotational symmetry? If so, state the angles of rotation (less than 360°) which conserve the set of points.

11 Find k if the given point lies on the given circle in each case:

a $(k, 0)$, $x^2 + y^2 = 1$ *b* $(12, k)$, $x^2 + y^2 = 169$
c $(k, -2)$, $x^2 + y^2 = 13$ *d* (k, k), $x^2 + y^2 = 72$

12 Find the coordinates of the points of intersection of the circle $x^2 + y^2 = 25$ with each of the following lines, and, in each case state the length of the chord formed.

a $y = 3$ *b* $y = 4$ *c* $y = 4.9$ *d* $y = 5$ *e* $y = 5.1$

Example. A is the point $(1, 0)$ and B is $(9, 0)$. P(x, y) belongs to the locus given by $\{P: PB = 3PA\}$. Show that the locus of P is the circle with equation $x^2 + y^2 = 9$.

$\{P: PB = 3PA\}$

$= \{P: \sqrt{(x-9)^2 + (y-0)^2} = 3\sqrt{(x-1)^2 + (y-0)^2}\}$

$= \{P: x^2 - 18x + 81 + y^2 = 9(x^2 - 2x + 1 + y^2)\}$

$= \{P: x^2 - 18x + 81 + y^2 = 9x^2 - 18x + 9 + 9y^2\}$

$= \{P: 8x^2 + 8y^2 = 72\}$

$= \{P: x^2 + y^2 = 9\}.$

The locus of P is the circle with equation $x^2 + y^2 = 9$, having centre the origin and radius 3 units, as shown in Figure 17.

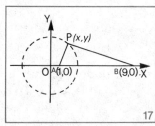

Exercise 6B

1   O is the origin and M is the point (6, 0). Show that the equation of the locus of P(x, y) given by {P: PO = PM} is the straight line with equation $x = 3$. Illustrate with a sketch.

2   O is the origin and N is the point (-8, 0). Show that the equation of the locus of P(x, y) given by {P: PO = PN} is the straight line with equation $x = -4$.

3   A is the point (1, 0) and B is (4, 0). Show that the equation of the locus of P(x, y) given by {P: PB = 2PA} is a circle with centre the origin, and state its radius.

4   Repeat question *3* for A(-1, 0), B(-9, 0) and {P: PB = 3PA}.

5   Repeat question *3* for A(0, 2), B(0, 8) and {P: PB = 2PA}.

6   Repeat question *3* for A(0, -1), B(0, -25} and {P: PB = 5PA}.

7   Repeat question *3* for A(1, 1), B(9, 9) and {P: PB = 3PA}.

8   A is the point (-3, 4) and B is (2, -1). Show that the equation of the locus of P(x, y) given by {P: 2AP = 3PB} is $x^2 + y^2 - 12x + 10y - 11 = 0$.

9   O is the origin, and the foot of the perpendicular from P(x, y) to the line with equation $x = -2$ is M. Show that the equation of the locus of P given by {P: PO = PM} is $y^2 = 4(x + 1)$ (a parabola).

10   If the foot of the perpendicular from P(x, y) to the line with equation $y = -5$ is N, show that the equation of the locus of P given by {P: PO = PN} is $x^2 = 5(2y + 5)$.

Summary

1 a Every diameter of a circle is an axis of *bilateral symmetry* of the circle.

b A diameter (radius) perpendicular to a chord bisects the chord.

c A diameter (radius) which bisects a chord is perpendicular to it.

d The perpendicular bisector of a chord passes through the centre of the circle.

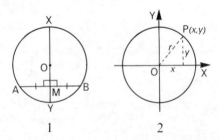

1   2

2 *The equation of the circle, centre O and radius r, is*

$$x^2 + y^2 = r^2$$

3 Dilatation

1 Enlarging and reducing shapes

In the chapter on similar shapes in Book 5 we saw that two shapes are similar if
(i)   they are equiangular, *and*
(ii) their corresponding sides are in proportion.
   In this chapter we first use these ideas to investigate the problem of enlarging or reducing a given shape.

Exercise 1

1   Look at the 'spider's web' diagram in Figure 1.

a Why is angle $OP_1 Q_1 =$ angle $OP_2 Q_2$?
b Why are the two hexagons $P_1 Q_1 R_1 S_1 T_1 U_1$ and $P_2 Q_2 R_2 S_2 T_2 U_2$ equiangular?
c State the numerical values of

$$\frac{P_2 Q_2}{P_1 Q_1}, \quad \frac{Q_2 R_2}{Q_1 R_1}, \quad \frac{R_2 S_2}{R_1 S_1}, \quad \frac{S_2 T_2}{S_1 T_1}, \quad \frac{T_2 U_2}{T_1 U_1} \quad \text{and} \quad \frac{U_2 P_2}{U_1 P_1}.$$

d Why can we say that the two hexagons are similar?

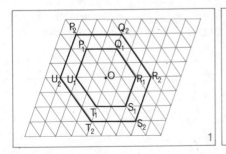

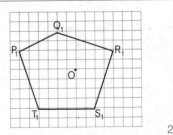

127

Note. Since the hexagons are similar we can regard each as a scale drawing, either *enlarged* or *reduced*, of the other. We can make such scale drawings easily as shown in question *2*.

2　In Figure 2, we are given the centre of the 'web' O, and a pentagon $P_1 Q_1 R_1 S_1 T_1$. We wish to draw a *similar* pentagon with sides half as long as those of $P_1 Q_1 R_1 S_1 T_1$.

　　a　Make a copy of the pentagon $P_1 Q_1 R_1 S_1 T_1$ on squared paper, and mark the point O.

　　b　Join O to each vertex of the pentagon.

　　c　Mark P_2 the midpoint of OP_1, Q_2 of OQ_1, R_2 of OR_1, S_2 of OS_1 and T_2 of OT_1.

　　d　Draw the pentagon $P_2 Q_2 R_2 S_2 T_2$.

　　e　Name pairs of triangles in the completed figure that you expect to be similar.

　　f　Name pairs of parallel lines in the completed figure.

3　Start again with Figure 2, but this time construct a pentagon with sides twice as long as those of $P_1 Q_1 R_1 S_1 T_1$.

4　Repeat question *3* for a different position of O inside $P_1 Q_1 R_1 S_1 T_1$. What can you say about the size and shape of the two enlargements?

We call O the *centre* and 2 the *scale factor* of the *enlargement*. In question *2*, the centre is O and the scale factor of the *reduction* is $\frac{1}{2}$. We can describe the enlargement or reduction by the notation [O, *k*] where O is the *centre* and *k* is the *scale factor*.

Note that in Figure 1 the scale factor is given by

$$\frac{P_2 Q_2}{P_1 Q_1} \quad \text{or by} \quad \frac{OP_2}{OP_1}.$$

5　Copy the shape in Figure 3 (i) on squared paper. With centre O and scale factor 2 construct an enlargement $A_1 B_1 C_1 D_1$ as follows.

　　a　Join OA and produce it to A_1 so that

$$\frac{OA_1}{OA} = 2, \quad \text{i.e. } OA_1 = 2OA.$$

　　b　Do the same for OB, OC and OD, and draw the square $A_1 B_1 C_1 D_1$.

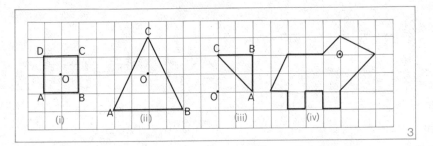

6   Copy the shape in Figure 3 (ii) on squared paper and draw a reduction of triangle ABC using centre O and scale factor $\frac{1}{2}$.

7   Copy the shape in Figure 3 (iii) on squared paper and draw an enlargement of triangle ABC using centre O and scale factor 3.

8   Copy the shape in Figure 3 (iv) on squared paper and draw its image under the enlargement [O, 2].

9   Using ruler, pencil and blank paper, draw the enlargements or reductions of the following shapes with the given centres and scale factors.

   a   square ABCD; centre O inside the square; scale factor 3
   b   square PQRS; centre O outside the square; scale factor 2
   c   rectangle WXYZ; centre W; scale factor $\frac{1}{2}$
   d   kite EFGH; centre O inside the kite; scale factor $\frac{3}{2}$
   e   parallelogram KLMN; centre O the midpoint of KL; scale factor $\frac{1}{2}$.

10   In each diagram in Figure 4 the original shape is drawn in solid lines and the enlargement or reduction in broken lines. Write down the scale factor for each diagram and say whether there has been an enlargement or a reduction.

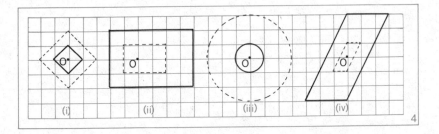

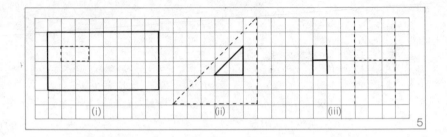

11 Copy the diagrams in Figure 5 on squared paper and in each case find the position of the centre of enlargement or reduction and the scale factor.

12 a Draw a triangle ABC on squared paper. Construct triangle $A_1 B_1 C_1$ with its sides parallel to those of triangle ABC and half the lengths of those of triangle ABC.

b On the same diagram construct triangle $A_2 B_2 C_2$ with its sides parallel to those of triangle ABC and double the lengths of those of triangle ABC.

c In what ways are the lengths and directions of the sides of triangles $A_1 B_1 C_1$ and $A_2 B_2 C_2$ related to each other?

13 On squared paper draw a shape with straight or curved sides, choose a centre and scale factor, and sketch several enlargements and reductions of the shape.

2 $\overrightarrow{OP'} = k\overrightarrow{OP}$ (*k positive*)

In Section 1 we obtained enlargements and reductions by choosing a centre O and a scale factor k. For a point P in the given shape we obtained a corresponding point P' such that

(i) O, P and P' were collinear, and

(ii) $\overrightarrow{OP'} = k \overrightarrow{OP}$.

Using directed line segments we can replace (i) and (ii) by the relation $\overrightarrow{OP'} = k\overrightarrow{OP}$. For example in Figure 6 (i) $\overrightarrow{OP'} = 3 \overrightarrow{OP}$, and in Figure 6 (ii) $\overrightarrow{OP'} = \frac{1}{4} \overrightarrow{OP}$.

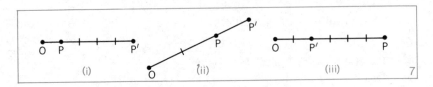

Exercise 2

1   For each diagram in Figure 7, write down a relation $\overrightarrow{OP'} = k\ \overrightarrow{OP}$, putting in the appropriate value of k.

2   Draw a line OP and for each of the following scale factors (k) find the position of P' such that $\overrightarrow{OP'} = k\ \overrightarrow{OP}$.

a 2   *b* 4   *c* $\frac{1}{2}$ *d* $\frac{3}{2}$ *e* 1

In Figure 8, P is the point (2, 1), so $\overrightarrow{OP}$ represents the vector $\begin{pmatrix} 2 \\ 1 \end{pmatrix}$.

For scale factor 3, $\overrightarrow{OP'} = 3\ \overrightarrow{OP}$.

Hence $\overrightarrow{OP'}$ represents the vector $3\begin{pmatrix} 2 \\ 1 \end{pmatrix} = \begin{pmatrix} 6 \\ 3 \end{pmatrix}$,

and P' is the point (6, 3).

In the same way, if Q is the point (2, 3), Q' is the point (6, 9).

In general, under an enlargement or reduction [O, k] a point P(a, b)→P'(ka, kb).

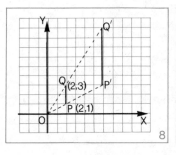

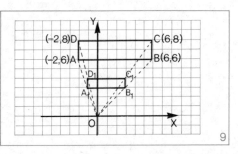

3 In Figure 9, the coordinates of A, B, C and D are shown. $A_1 B_1 C_1 D_1$ is obtained from ABCD by the reduction $[O, \frac{1}{2}]$. Find the coordinates of A_1, B_1, C_1 and D_1.

4 Repeat question 3 for the enlargement $[O, \frac{3}{2}]$

5 Draw triangle OAB, where O is the origin, A is the point (4, 0) and B is (0, −2). Show triangle $OA_1 B_1$ obtained by the enlargement $[O, 2]$ and triangle $OA_2 B_2$ obtained by the reduction $[O, \frac{1}{4}]$. What are the coordinates of A_1, B_1, A_2 and B_2?

6 The coordinates of the vertices of a triangle are A(2, 2), B(6, 4) and C(4, 8).

  a Plot these points on squared paper.
  b Draw $A_1 B_1 C_1$ which is a reduction $[O, \frac{1}{2}]$ of triangle ABC.
  c Draw $A_2 B_2 C_2$ which is an enlargement $[O, \frac{5}{4}]$ of triangle ABC.
  d Write down the coordinates of A_1, B_1, C_1 and A_2, B_2, C_2.
  e State any connection you see between the coordinates of A, B, C and those of A_1, B_1, C_1 and A_2, B_2, C_2.

7 a Draw the triangle of question 6 on squared paper.
  b Draw the triangle $A_1 B_1 C_1$ which is an enlargement $[O, \frac{3}{2}]$ of triangle ABC.
  c Draw the triangle $A_2 B_2 C_2$ which is a reduction $[O, \frac{1}{2}]$ of triangle $A_1 B_1 C_1$.
  d Write down the coordinates of A_1, B_1, C_1 and A_2, B_2, C_2.
  e What reduction would map triangle ABC to triangle $A_2 B_2 C_2$?

8 Sketch a triangle OAB and its enlargement OA′B′ for centre O and scale factor 4. Copy and complete the following:

$$\overrightarrow{OA'} = 4\overrightarrow{OA} \quad \text{and} \quad \overrightarrow{OB'} = ...\overrightarrow{OB}.$$
$$\overrightarrow{A'B'} = \overrightarrow{OB'} - \overrightarrow{OA'}$$
$$= 4\overrightarrow{OB} - ...\overrightarrow{OA}$$
$$= 4(... - ...)$$
$$= 4\overrightarrow{AB}.$$

What does this result tell us about the lengths and directions of the line segments A′B′ and AB?

9 Repeat question 8 for the scale factor $\frac{1}{2}$ to prove that $\overrightarrow{A'B'} = \frac{1}{2}\overrightarrow{AB}$.

10   Repeat question *8* for the scale factor k to prove that $\overrightarrow{A'B'} = k\,\overrightarrow{AB}$.

11   Under an enlargement with centre O, A→A′ and B→B′. Find the scale factor of the enlargement for each of the following cases:

a $\overrightarrow{OA'} = 2\overrightarrow{OA}$ b $\overrightarrow{OA} = \frac{1}{3}\overrightarrow{OA'}$ c $\overrightarrow{OB'} = \frac{4}{3}\overrightarrow{OB}$ d $\overrightarrow{OB} = \frac{3}{4}\overrightarrow{OB'}$

e $\overrightarrow{A'B'} = 2\overrightarrow{AB}$ f $\overrightarrow{AB} = \frac{2}{3}\overrightarrow{A'B'}$ g $\overrightarrow{OA} = 2\overrightarrow{AA'}$ h $\overrightarrow{BB'} = \frac{1}{5}\overrightarrow{OB}$

---

3 $\overrightarrow{OP'} = k\,\overrightarrow{OP}$ (*k negative*)

In Figure 10(i) the directions of the line segments OP and OP′ are opposite, so the relation must be $\overrightarrow{OP'} = -2\overrightarrow{OP}$, i.e. $k = -2$.

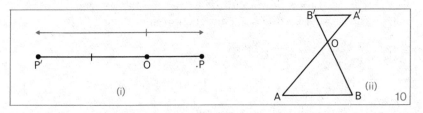

(i)

(ii)

10

In Figure 10(ii), $k = -\frac{1}{2}$ and we have $\overrightarrow{OA'} = -\frac{1}{2}\overrightarrow{OA}$ and $\overrightarrow{OB'} = -\frac{1}{2}\overrightarrow{OB}$.

Exercise 3

1   Start by drawing a line OP for each part of this question. Find the position of P′ such that $\overrightarrow{OP'} = k\overrightarrow{OP}$ when k is replaced by:

a -2 b -1 c $-\frac{1}{2}$ d $-\frac{1}{4}$ e -5.

2   Copy Figure 11 on squared paper.

a Plot the position of points A_1, B_1 and C_1 such that $\overrightarrow{OA_1} = -\overrightarrow{OA}$, $\overrightarrow{OB_1} = -\overrightarrow{OB}$ and $\overrightarrow{OC_1} = -\overrightarrow{OC}$. What is the scale factor here?

b Which of the following are true?

   (*1*) $\overrightarrow{A_1B_1} = \overrightarrow{AB}$ (*2*) $\overrightarrow{B_1C_1} = -\overrightarrow{BC}$ (*3*) $\overrightarrow{C_1A_1} = \overrightarrow{AC}$

3   Repeat question *2a* for $\overrightarrow{OA_1} = -2\overrightarrow{OA}$, $\overrightarrow{OB_1} = -2\overrightarrow{OB}$ and $\overrightarrow{OC_1} = -2\overrightarrow{OC}$.

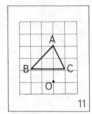

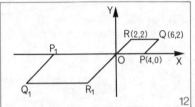

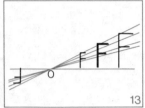

4   In Figure 12, $OP_1 Q_1 R_1$ is obtained from OPQR by means of the enlargement $[O, -\frac{3}{2}]$. Find the coordinates of P_1, Q_1 and R_1.

5   Repeat question *4* for the reduction $[O, -\frac{1}{2}]$. Illustrate by a diagram.

6   Figure 13 shows how a given **F**-shape can be enlarged, reduced or turned round by drawing construction lines through O.

a   Copy Figure 13 on plain paper.

b   Draw sketches for different capital letters or numerals.

7   The vertices of a triangle are $A(-5, 0)$, $B(-2, 0)$ and $C(3, 4)$.

a   Plot these points on squared paper.

b   Draw triangle $A_1 B_1 C_1$ which is the image of triangle ABC under $[O, -1]$, and state the coordinates of A_1, B_1 and C_1.

c   What other transformations would produce the same image?

8   The vertices of a rhombus are $P(-1, -1)$, $Q(3, 1)$, $R(5, 5)$ and $S(1, 3)$.

a   Draw the rhombus PQRS on squared paper.

b   On the same diagram draw $P_1 Q_1 R_1 S_1$, the image of PQRS under the enlargement $[O, -2]$.

c   State the coordinates of P_1, Q_1, R_1 and S_1.

9    $A(-1, 1)$, $B(3, 1)$, $C(3, 3)$ and $D(-1, 3)$ are the vertices of a rectangle. P is the point $(2, 0)$.

On squared paper draw ABCD and its image $A_1 B_1 C_1 D_1$ under the enlargement $[P, -3]$, and write down the coordinates of A_1, B_1, C_1 and D_1.

Why, in this case, can you not multiply the original coordinates by -3?

4 Dilatation: a transformation of the plane

We have already met, more than once, the idea of a transformation of the plane (see Figure 14).

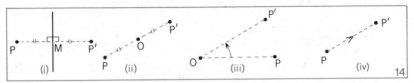

(i) *Reflection in a line* is a one-to-one mapping of the set of points of the plane onto itself.
(ii) So is *reflection in a point.*
(iii) So is *rotation about a point* through a given angle.
(iv) So is *translation.*

Now, corresponding to the activity of enlarging and reducing figures of Sections 1–3, we can think of a transformation of the plane under which, to each point P of the plane, there corresponds one point P′ such that $\overrightarrow{OP'} = k\overrightarrow{OP}$, where O is a given fixed point and k is a given fixed number.

The name given to this transformation is *dilatation.* We sometimes indicate the *centre* and the *scale factor* of the dilatation by referring to 'the dilatation $[O, k]$' or 'the dilatation $[P, -\frac{3}{4}]$', etc.

A major difference between dilatation and the transformations shown in Figure 14 is that, if A→A′ and B→B′ under a reflection or a rotation or a translation, then the distance A′B′ is equal to the distance AB while under a dilatation, since $\overrightarrow{A'B'} = k\ \overrightarrow{AB}$, the distance A′B′ equals the distance AB only if $k = 1$ or -1.

We say that under a dilatation, distance is *not conserved.*

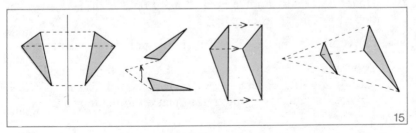

Figure 15 illustrates that under a reflection, a rotation, or a translation, a figure and its image are congruent, but under a dilatation they are similar.

Example. Under the dilatation $[O, 4]$ find the images of the points (i) $A(4, 3)$ and (ii) $D(r, s)$.

(i)   Under this dilatation suppose that $A \rightarrow A'(a', b')$.

$$\text{Then} \quad \overrightarrow{OA'} = 4\overrightarrow{OA} \Leftrightarrow \begin{pmatrix} a' \\ b' \end{pmatrix} = 4\begin{pmatrix} 4 \\ 3 \end{pmatrix} = \begin{pmatrix} 16 \\ 12 \end{pmatrix}.$$

A' is the point $(16, 12)$.

(ii)   If $D(r, s) \rightarrow D'(r', s')$, then $\overrightarrow{OD'} = 4\overrightarrow{OD}$

$$\Leftrightarrow \begin{pmatrix} r' \\ s' \end{pmatrix} = 4\begin{pmatrix} r \\ s \end{pmatrix} = \begin{pmatrix} 4r \\ 4s \end{pmatrix}.$$

D' is the point $(4r, 4s)$.

Note that we can find the coordinates of the image if we multiply by the scale factor, *provided the centre of dilatation is the origin.*

Exercise 4

1   In Figure 14, describe in each case how you would find the image of P.

2   Under a dilatation a line and its image are parallel.

  a Is this always true for:
    (1) a translation   *(2)* a reflection   *(3)* a rotation?
  b In the cases where it is not always true, state or show in diagrams the special cases in which it is true.

3   Under a transformation a point P maps onto itself. State the position of P if the transformation is:

  a reflection in a line   *b* reflection in a point   *c* rotation

4   Copy and complete the following table by putting a tick in each place where the property in the left-hand column is *conserved* (i.e. remains unchanged) by the transformation in the top row.

	Reflection	Rotation	Translation	Dilatation ($k \neq 1$)	
				$k > 0$	$k < 0$
Length of line segment					
Direction of line					
Size of angle					
Area of triangle					

5   Find the images of the points A(4, 2), B(-2, 3), C(-1, 0) and D(-2, -1) under the dilatations:

    a   [O, 5]   *b*   [O, -3]   *c*   [O, $-\frac{3}{2}$].

6   Plot the points A(4, 8), B(8, 8), C(8, -2) and D(4, -2), and on the same diagram plot the images of these points under the dilatation [O, $-\frac{1}{2}$].

7 *a*   On squared paper draw the square O(0, 0), A(1, 0), B(1, 1), C(0, 1).
   b   Sketch the image squares under dilatations [O, 8], [O, 4], [O, 2], [O, 1], [O, $\frac{1}{2}$], [O, $\frac{1}{4}$], [O, $\frac{1}{8}$].
   c   Starting with the image under [O, 8] write down the sequences of:
      (*1*)   coordinates of images of B   (*2*)   areas of image squares.

8   C(2, 1) is the centre of the dilatations in this question.

   a   On squared paper draw the images of the rectangle with vertices P(1, 0), Q(5, 0), R(5, 3), S(1, 3) under the dilatations:
      (*1*)   [C, 2].   (*2*)   [C, -1].
   b   What can you state about all three rectangles?

9 *a*   A is (4, 1), B is (7, 3), C is (3, 9). Calculate the lengths of the sides of triangle ABC, and prove that it is right-angled.
   b   Find the images A′, B′, C′ of A, B, C under the dilatation [O, 2].
   c   Calculate the lengths of the sides of triangle A′B′C′, and prove that it is right-angled.

10*a*   List the coordinates of six points on the straight line $y = 2x$.
   b   Find the coordinates of their images under the dilatation [O, 3].
   c   State the equation of the straight line through these image points.
   d   Illustrate by means of a sketch on squared paper.

11   Repeat question *10* for the line with equation $y = -4x$.

12   Repeat question *10* for the line with equation $x = 3$.

13*a*   List the coordinates of six points on the line $y = 2x + 4$.
   b   List the coordinates of their images under the dilatation [O, 3].

c Write down the equation of the line through these images.

d Illustrate by means of a sketch on squared paper.

14 Repeat question *13* for the dilatations:

a $[O, 2]$ *b* $[O, -\frac{1}{2}]$.

15 Look at questions *10* and *11* again, and write down the equation of the image of the line $y = mx$ under the dilatation $[O, 3]$.

16 Look at questions *13* and *14* again, and write down the equation of the image of the line $y = 2x + 4$ under the dilatation $[O, k]$.

17a Draw the lines $x = 2$, $y = 4$ and $x + y = 8$, and shade the region for which $x \geqslant 2$, $y \geqslant 4$ and $x + y \geqslant 8$.

b Find the images of the lines in *a* under the dilatation $[O, \frac{3}{2}]$, and shade the image of the corresponding region.

Exercise 4B

1 a List the coordinates of five points on the parabola $y = x^2$.

b List the coordinates of their images under the dilatation $[O, 2]$.

c Check that the coordinates of their images satisfy the equation $y = \frac{1}{2}x^2$.

2 Repeat question *1 a* and *b*, for the dilatations:

a $[O, 4]$ *b* $[O, -2]$.

In each case write down the equation of the image parabola.

3 Look at your answers to questions *1* and *2* and write down the equation of the image of $y = x^2$ under the dilatation $[O, k]$.

4 On squared paper mark the vertices of a square ABCD as given below. Under a dilatation $[M, 2]$ the images of A, B, C and D are P, Q, R and S respectively. Find the coordinates of P, Q, R and S in each case.

	M	A	B	C	D
a	(5, 1)	(8, 2)	(9, 2)	(9, 3)	(8, 3)
b	(0, 1)	(−1, 0)	(1, 0)	(1, 2)	(−1, 2)
c	(5, 3)	(3, 2)	(6, 2)	(6, 5)	(3, 5)

5 For each part of question *4* work out $x_A - x_M$ and $x_P - x_M$; also $x_C - x_M$ and $x_R - x_M$; also $y_B - y_M$ and $y_Q - y_M$. Comment on the results.

6 a Plot the points A(3, 6), B(1, 3) and C(4, 2).

b If A'B'C' is the image of △ABC under a dilatation with scale factor 3, and A' is (9, 8), find the centre of the dilatation and the co-ordinates of B' and C'.

c Repeat b for the image A"B"C" of △ABC under a dilatation with scale factor −1, and A" (1, 2).

(Note that both images are not only similar to the original figure but have their sides parallel to the corresponding sides of the original figure. This happens in all dilatations. In such a situation we describe the figures as *similar* and *similarly situated*.)

7 a Plot the points A(9, 9), B(11, 9), C(11, 6), D(9, 6), P(17, 9), Q(21, 9), R(21, 3) and S(17, 3).

b If P, Q, R and S are images of A, B, C and D under a dilatation, find its centre and scale factor.

c Find the centre and scale factor of the dilatation under which P, Q, R and S are the images of C, D, A and B respectively.

8 In Figure 16, ABC and A'B'C' are similar triangles with corresponding sides parallel. A'A and B'B produced meet at O. A'B' = kAB.

a Copy and complete the following.

$$\overrightarrow{OC'} = ... + \overrightarrow{A'C'} = k\,\overrightarrow{OA} + ... = k(\overrightarrow{OA} + ...) = k\,\overrightarrow{OC}.$$

b State what information this gives you about OC and OC'.

9 In triangle ABC, B'C' is parallel to BC, and AM is a median (see Figure 17). Prove that AM bisects B'C'. (*Hint.* Consider a dilatation, centre A, such that B→B'.)

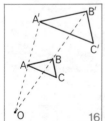 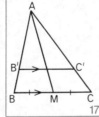 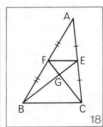 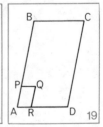

10 In Figure 18, BE and CF are medians of triangle ABC.

a Prove that FE is parallel to BC and equal to half of it. (*Hint.* Use [A, ½].)

b Prove that BE and CF divide each other in the ratio 2:1.
(*Hint.* Use [G, −2].)

c Use *a* and *b* to prove that the medians of a triangle trisect one another.

11 In Figure 19, ABCD and APQR are parallelograms in which $AP = \frac{1}{4}AB$ and $AR = \frac{1}{4}AD$. Prove that A, Q and C are collinear.

Summary

1   Under a *dilatation* a figure and its image are similar and corresponding sides are parallel.

2   *A dilatation, centre O, and scale factor k* is written $[O, k]$.

3   If under a dilatation $[O, k]$, A→A′ and B→B′, we have

(i)

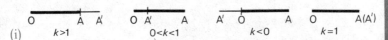

(ii) $\overrightarrow{OA'} = k\,\overrightarrow{OA}$, $\overrightarrow{OB'} = k\,\overrightarrow{OB}$ and $\overrightarrow{A'B'} = k\,\overrightarrow{AB}$.

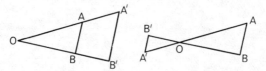

(iii)   If the origin is the centre of dilatation and k is the scale factor, $(a, b) \rightarrow (ka, kb)$.

Topics to explore

(i) The circumference of the earth

Figure 1 shows how Eratosthenes measured the circumference of the earth in the 3rd century B.C. When the sun was directly overhead at S (Syene or Aswan) he measured the size of the shaded angle at A (Alexandria) to be $7\frac{1}{2}°$. He then deduced the size of the shaded angle at C.

Knowing that the arc joining A to S was 800 km long, he was able to calculate that the circumference of the earth was about 38 400 km. Can you?

Can you find out how he measured the shaded angle at A?

Can you find out how the Egyptians knew the distance between Alexandria and Syene?

Why are the shaded angles equal?

What must Eratosthenes have believed about the shape of the earth?

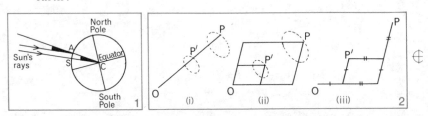

(ii) The pantograph

Using what we know about enlarging or reducing we can make an instrument called a *pantograph* to enlarge or reduce shapes.

Figure 2 (i) illustrates how a piece of elastic fixed at O can be used as a very simple pantograph. As P traces out a shape, P′ traces out a reduced copy. As the elastic must be kept taut all the time, this is not a very adaptable instrument.

Figure 2 (ii) shows a system of jointed rods forming two *similar* parallelograms with O fixed. Again P and P′ trace out similar figures.

Figure 2 (iii) shows a pantograph requiring less material. Equal rods are indicated in the diagram; again O is fixed.

In each case if P traces out a figure, P' traces out a reduced copy. We could quite easily reverse the process by making P' trace a figure; then P traces out an enlarged copy.

Can you construct a pantograph for yourself? Can you think of a way of fixing a pencil to trace the required copy? Can you construct a pantograph to enlarge or reduce in any given ratio?

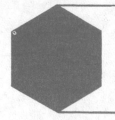
Revision Exercise on Chapter 1
Circle 1. Rotational Symmetry

Revision Exercise 1

1 In Figure 1, O is the centre of the circle. In this figure name:

a a diameter   *b* a radius   *c* a chord

d a minor arc   *e* a major arc.

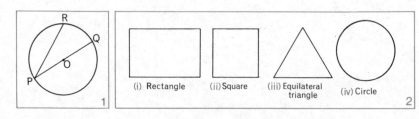

(i) Rectangle   (ii) Square   (iii) Equilateral triangle   (iv) Circle

2 Write down the order of rotational symmetry of each shape in Figure 2.

3 The vertices of △ABC are A(3, 1), B(1, 6) and C(−2, 4). Plot these points on squared paper and draw the triangle. On the same diagram draw the image △A′B′C′ of △ABC under an anticlockwise (positive) rotation of 90° about the origin. State the coordinates of A′, B′ and C′.

4 What is the least positive angle through which an equilateral triangle must be rotated about its centre so as to fit its outline?

5 Figure 3 shows a circle with nine equally spaced radii.

144

a   Under an anticlockwise rotation of 40° about O, to what do the following map?

(*1*)   **B**   (*2*)   arc DE   (*3*)   chord FG

(*4*)   sector HOK   (*5*)   triangle AOC.

b   If the polygon ABCDEFGHK is completed, what can you say about:

(*1*)   all its sides   (*2*)   the size of each of its angles?

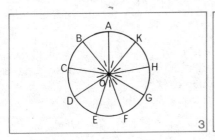

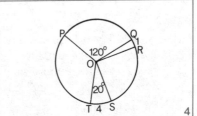

6   In Figure 4, O is the centre of a circle in which ∠TOS = 20°, ∠QOP = 120°, arc TS = 4 cm and arc RQ = 1 cm. Calculate:

a   the length of arc PQ   *b*   the size of ∠QOR
c   the ratio of the area of sector TOS to the area of the circle.

7   The tip of the minute hand of a clock moves round the circumference of a circle on which the minute positions are marked.

a   The length of the arc joining '10 past' and '20 past' is 2 cm. Calculate the length of the arc joining '5 past' and '25 to'.
b   The area of the sector swept out by the minute hand from 'half past' to 'quarter to' is 15 cm². Calculate the area swept out from '10 past' to '19 past'.

8   In a circle with centre O and radius 18 cm, an arc PQ subtends an angle of 50° at the centre. Calculate:

a   the length of arc PQ
b   the area of sector OPQ.   (Take $\pi = 3.14$.)

9   In Figure 5, AB and BC are equal chords in a circle with centre O. If ∠OAB = 62°, calculate the size of ∠AOC.

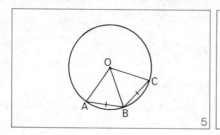
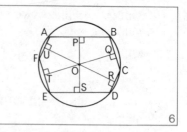

10 In Figure 6, pairs of opposite sides of hexagon ABCDEF are equal. O is the centre of the circle, and OP, OQ, OR, OS, OT, OU are perpendicular to the sides of the hexagon. Name pairs of equal lines drawn from O.

11a PQ and RS are equal and parallel chords in a circle, centre O. Draw a diagram and explain how to find an axis of symmetry of the diagram. Is there more than one answer?

   b Repeat *a* when PQ and RS are equal but not parallel.

12 A regular 20-sided polygon is inscribed in a circle. Calculate:

   a the size of each angle at the centre subtended by a side of the polygon

   b the size of each angle of the polygon.

13 Construct a regular pentagon. Produce each side of the pentagon both ways to form a five-pointed star. Calculate the size of each angle at the points of the star.

14 A regular polygon has an angle of 170°. By considering the isosceles triangles which form the polygon, calculate:

   a the sizes of the angles in the triangles

   b the number of sides in the polygon.

15 Two arcs AB and CD of a circle, centre O, measure $2x$ cm and $5x$ cm respectively. If the area of sector OCD is $22\frac{1}{2}$ cm², calculate the area of sector OAB.

16 Construct a regular dodecagon with each of its 12 sides of length 3 cm.

17 Prove that the size of each angle of a regular polygon with n sides is

$$180° - \frac{360°}{n}.$$

Revision Exercise on Chapter 2
Circle 2. Bilateral Symmetry

Revision Exercise 2

1   Write down the number of axes of bilateral symmetry for each shape in Figure 2 on page 144.

2   Show how you would draw an axis of symmetry, given:

  a two intersecting circles   *b* two concentric circles.

3   In Figure 7, AOB is a diameter and CD is a chord cutting AB at P. Show, and explain, how you would draw a chord EF so that the whole figure has AB as an axis of symmetry. Name three pairs of equal lines in the completed diagram.

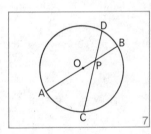 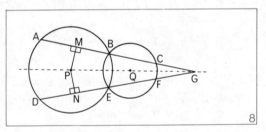

4   In Figure 8, P and Q are the centres of the two circles. ABC and DEF are double chords, which meet at G on PQ produced. PM and PN are perpendicular to AB and DE. Name:

  a the axis of symmetry of the figure
  b four pairs of equal lines.

5   State which of the following are true and which are false.

  a Every triangle has a circumcircle.
  b Given three points it is always possible to draw a circle through them.
  c The centre of a circle through P and Q lies on the perpendicular bisector of PQ.

d In a right-angled triangle the centre of the circumcircle is the mid-point of the hypotenuse.

6 In Figure 9, T is the centre of a circle of radius 2 cm. PA = 3·2 cm, and the figure is symmetrical about SR. Calculate:

a the length of TR   *b*   the length of RS
c the area of triangle PSA.

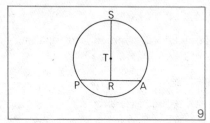
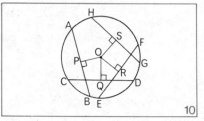

7 In Figure 10, O is the centre of a circle of radius 5 cm. AB = CD = EF = GH = 8 cm.

a Calculate the length of OP, OQ, OR and OS.
b Explain why P, Q, R and S lie on a circle, and state the centre and radius of the circle.

8 In a circle of radius 13 cm, chords of length 24 cm, 12 cm and 6 cm are drawn.

a Calculate the distance from the centre to each chord.
b Is the distance from the centre to a chord inversely proportional to the length of the chord?

9 Show the following sets of points on squared paper.

a $A = \{(x, y): x^2 + y^2 = 16\}$ *b* $B = \{(x, y): x < 3\}$ *c* $C = A \cap B$

10 Show, by shading, the region of the coordinate plane defined by

$\{(x, y): 4 \leqslant x^2 + y^2 \leqslant 25\}$.

11 Find where possible, the coordinates of the points of intersection of the circle $x^2 + y^2 = 100$ with each of the following lines. In each case state the length of the chord of the circle.

a $y = 8$ *b* $x = -8$ *c* $y = -6$ *d* $x = 10$ *e* $y = 12$

12 With the aid of a diagram describe the locus of the midpoints of equal chords in a circle, centre O.

13 In Figure 11, the two circles with centres X and Y intersect at C and
D. The radii of the circles are 5 cm and 10 cm respectively. If
AB = CD = RS, and AB is 4 cm from X, calculate the distance of RS
from Y, to one decimal place.

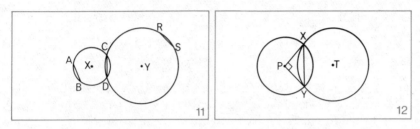

14 In Figure 12, P and T are the centres of the circles, and the radius of
the circle, centre T, is 10 cm. The common chord XY is 8 cm from T.
If ∠XPY = 90°, calculate:

a the length of XY
b the radius of the circle, centre P, to two significant figures.

15 An isosceles triangle PQR, with PQ = PR, is inscribed in a circle of
radius 17 cm. If QR = 30 cm, calculate the lengths of the altitude PS
and of the side PQ, to two significant figures.

16 Figure 13 shows a section through a horizontal pipe. Water in the
pipe is 18 cm deep at the middle of the section, and the width of the
water surface is 48 cm. Calculate the radius of the pipe, *r* cm.
(*Hint*. Express OC in terms of *r* and use Pythagoras' theorem.)

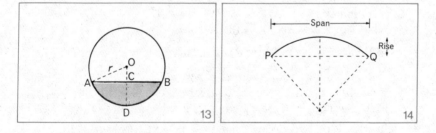

17 Figure 14 shows the 'span' and 'rise' of the circular arc PQ. If the
span is 20 m and the rise 1 m, calculate the radius of the circle.

18 ABCD is a rectangle in which AB = 42 cm and AD = 9 cm. The largest possible arc of a circle of which AB is a chord has to be drawn so that no part of the arc lies outside the rectangle. Calculate the radius of the circle.

19 Show the following sets of points on Cartesian diagrams.

a $S = \{(x, y): x^2 + y^2 \geqslant 100\}$ *b* $T = \{(x, y): x = 8\}$

c $S \cap T$ *d* $S \cup T$

20 The centre of the circumcircle of a triangle may lie inside the triangle, or on a side of the triangle, or outside the triangle. Explain, with the aid of sketches, how these three situations can arise.

Revision Exercise on Chapter 3
Dilatation

Revision Exercise 3

1 The vertices of triangle ABC are A(3, 1), B(0, 3) and C(−2, −2).

a Draw the triangle on squared paper.
b Draw an enlargement $A_1 B_1 C_1$ of triangle ABC with centre O, the origin, and scale factor 2. Give the coordinates of A_1, B_1 and C_1.
c What can you state about the two triangles?

2 The vertices of a kite OPQR are O(0, 0), P(6, 3), Q(6, 6) and R(3, 6).
a Draw the kite on squared paper.
b Draw the kite given by the reduction $[O, -\frac{1}{2}]$, and write down the coordinates of its vertices.
c What can you state about the two kites?

3 In Figure 15, rectangle PQRS is an enlargement of rectangle ABCD.

a Copy the diagram on squared paper.
b Find the centre and the scale factor of the enlargement.
c Find the centre and scale factor of the reduction of PQRS to ABCD.

4 In Figure 16, two circles C_1 and C_2 have their centres at the origin.

a State the centre and scale factor of the enlargement which transforms C_1 to C_2.

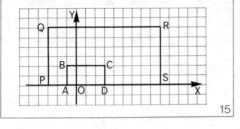

b State the centre and scale factor of the reduction which transforms C_2 to C_1.

c Describe the circle which is obtained from C_2 by the reduction $[O, \frac{2}{3}]$.

5 What are the images of the following points under a dilatation with centre the origin and scale factor 3?

a $(3, 2)$ *b* $(-1, 4)$ *c* $(2, -5)$ *d* $(0, 0)$

6 Repeat question 5 for a scale factor -2.

7 Which of the following statements are true and which are false?

a Under a dilatation, scale factor 2, the image of $(3, 2)$ is always $(6, 4)$.

b The dilatation $[O, -1]$ is the same transformation as a half turn about the origin.

c Under every dilatation, a figure and its image are:
(1) always similar   (2) always congruent
(3) never congruent.

d Under a dilatation, a polygon and its image have corresponding sides parallel.

e Given two squares, with their sides parallel, it is always possible to find a dilatation which maps one to the other.

8 Draw a rectangle ABCD with its diagonals intersecting at X. If a dilatation, centre X, maps B to D, state the points to which A, C and D are mapped by the same dilatation.

9 *a* List six points on the line with equation $y = 3x + 1$.

b List the coordinates of their images under the dilatation $[O, 2]$.

c Write down the equation of the line through the set of image points.

10 Sketch the lines with equations:

a $y = 4x + 2$ *b* $y = 4x + 5$ *c* $y = 4x - \frac{1}{3}$ *d* $y - 4x = 2$

What are the scale factors of the dilatations, centre O, which map lines *b*, *c* and *d* onto *a*?

11a Draw a circle C with centre (1, 1) and radius 1 unit, on squared paper.

  b Find the centre and radius of the image circle C_1 under a dilatation [O, 3].

  c Find the centre and radius of the circle C_2 which is the image of C_1 under a dilatation [P, 2], where P is the point $(-2, 0)$.

  d Find the centre and scale factor of the dilatation which would map C to C_2.

12 The position vectors of the vertices of a quadrilateral ABCD with respect to an origin O are *a*, *b*, *c*, *d* respectively.

  a If ABCD is mapped to $A_1 B_1 C_1 D_1$ under a dilatation, centre O and scale factor 2, write down the position vectors of A_1, B_1, C_1, D_1 in terms of *a*, *b*, *c*, *d*.

  b What can you deduce about AC and $A_1 C_1$?

  c Answer the same questions when the scale factor is -1.

13 Under a dilatation, the images of A(3, 1), B(6, 1), C(6, 7), D(3, 7) are P(7, 3), Q(8, 3), R(8, 5), S(7, 5) respectively. Find the centre and scale factor of the dilatation.

Cumulative Revision Section (Books 1-6)

Cumulative Revision Exercise

1   Make sketches of nets that could be folded to form:

a   a cube   *b*   a cuboid   *c*   a square pyramid.

2   A point A is 32 m from the foot of a tree on level ground. The angle of elevation of the top of the tree from A is 35°. By means of a scale drawing find the height of the tree to the nearest metre.

3   Euler's formula for solid figures states that $F + C = E + 2$ where F, C and E are the numbers of faces, corners and edges respectively. Verify that the formula is true by writing down the values of F, C and E for:

a   a cuboid   *b*   a triangular pyramid   *c*   a square pyramid
d   a cube with one corner cut off.

4   A rectangle has its sides parallel to rectangular axes. If one vertex is the point (2, 1) and the intersection of the diagonals is the point (5, 3), sketch the rectangle on squared paper and state the coordinates of the other three vertices.

5   The following sets of points are vertices of triangles. In each case plot the points and state the kind of triangle formed.

a   (3, 1),   (1, 3),   (−4, −4)   *b*   (−2, 1),   (−2, −4),   (3, 1)

6   A ship sails 6 km on a bearing of 135° and then 10 km on a bearing of 265°. By means of an accurate scale drawing find how far it is from its starting point.

7   If the universal set E is the set of quadrilaterals, draw a Venn diagram to illustrate the relationship between the sets of parallelograms, rectangles and squares.

8   Illustrate on squared paper the set of points

$$\{(x, y): -2 < x \leqslant 3\} \cap \{(x, y): 1 \leqslant y < 4\}.$$

9   Sketch all the different shapes that can be made by fitting together equal sides of:

a   two congruent right-angled triangles
b   two congruent triangles.

10   A ladder 4·1 m long standing on level ground reaches 4 m up a vertical wall. How far is the foot of the ladder from the wall?

11   Calculate the distances between the following pairs of points.

a   (2, 6) and (8, 14)   b   (−1, −11) and (−6, 1)
c   (7, 0) and (0, −1)

12   The diagonals of a rhombus measure 16 cm and 12 cm. Calculate:

a   the length of a side   b   the area.

13   In Figure 1, BM = MC. Complete each of the following equations by a directed line segment.

a $\overrightarrow{AB} + \overrightarrow{BC} = \ldots$ b $\overrightarrow{AB} + \frac{1}{2}\overrightarrow{BC} = \ldots$
c $\overrightarrow{AB} + \overrightarrow{BC} + \overrightarrow{CD} = \ldots$ d $\frac{1}{2}\overrightarrow{BC} + \overrightarrow{CD} = \ldots$

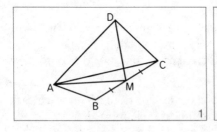
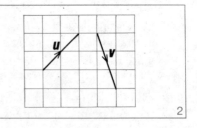

14   On squared paper copy the representatives of vectors **u** and **v** in Figure 2. Then draw a representative of each of the vectors:

a   **u** + **v**   b   **u** − **v**   c   **u** + 2**v**   d   **u** − 2**v**.

15   In Figure 3, AP = PQ = QB and $\overrightarrow{AB}$ represents the vector **u**. Write down in terms of **u** the vectors represented by:

a $\overrightarrow{AP}$ b $\overrightarrow{PB}$ c $\overrightarrow{AQ}$ d $\overrightarrow{BQ}$ e $\overrightarrow{QA}$. 、

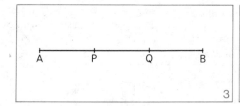

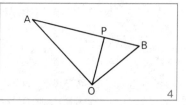

16   In Figure 4, the position vectors of A and B relative to the origin O
     are **a** and **b** respectively. Express the vector represented by $\overrightarrow{AB}$ in
     terms of **a** and **b**.
       If AP = 2PB, find in terms of **a** and **b** the vectors represented by:

   a $\overrightarrow{AP}$ b $\overrightarrow{PB}$ c $\overrightarrow{OP}$.

17   In each of the following equations find x in terms of **u** and **v**.

   a $u + x = 2u + v$ b $\frac{1}{3}(x + u) = \frac{1}{2}(u + v)$

18   In Figure 5 representatives of the vectors **a**, **b** and **c** are drawn.
     Express as number pairs the vectors:

   a **a**   b **3b**   c **−2c**   d **a+b**   e **a+2c**   f **2b−3c**.

       What is the magnitude of each of the vectors **a**, **b** and **c**?

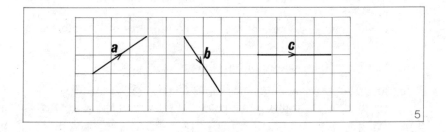

19   In triangles ABC and DEF, $\angle A = 63°$, $\angle B = 58°$, $\angle E = 58°$ and
     $\angle F = 59°$. Explain why the triangles are similar. Name pairs of
     corresponding sides.

20   In triangles ABC and PQR, AB = 4 cm, BC = 6 cm, CA = 8 cm,
     PQ = 12 cm, QR = 18 cm and PR = 24 cm. Explain why the
     triangles are similar, and name pairs of corresponding vertices.

21 In Figure 6: *a* are the shapes (i) and (ii) similar? *b* are the shapes (iii) and (iv) similar? In each case give reasons for your answer.

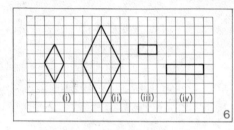

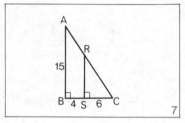

22 In Figure 7, AB and RS are perpendicular to BC. BS = 4 cm and SC = 6 cm. If AB = 15 cm, calculate the length of RS.

23 On a certain map the scale used is such that 1 mm on the map represents a distance of 100 000 mm on the earth's surface. What would be the length of a line on the map which represents a distance of 2 km on the ground?

24a Calculate, where possible, the gradients of the lines joining the points:
  (1) (4, 8) and (8, 4)   *(2)* (−6, 2) and (−6, −5)
  (3) (−2, 5) and (−6, 5).
  b Write down the equation of the line with gradient $\frac{1}{2}$ through the point (0, 3).
  c Write down the gradient, and the coordinates of the intersection with the *y*-axis, of the line with equation $y = -\frac{4}{3}x - 2$.

25 In Figure 8, the centre of the circle is O. Name:

  a an angle at the circumference subtended by arc TQ
  b the angle at the centre and an angle at the circumference subtended by chord QR.

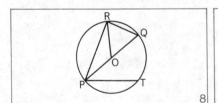

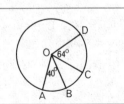

26 In Figure 9, O is the centre of the circle, and angles AOB and COD are 40° and 64° respectively.

a If the length of arc AB is 15 cm, calculate the length of arc CD.
b If the area of sector COD is 32 cm², calculate the area of sector AOB.

27 A circle has a radius of 10 cm. Calculate, to three significant figures, the length of an arc of this circle corresponding to an angle at the centre of 48°. (Take $\pi = 3\cdot14$.)

28 In Figure 10, T is the circumcentre of triangle ABC in which AB = 10 cm. The perpendicular distance from T to AB is 12 cm. Calculate: a the radius of the circle b the area of the circle. (Take $\pi = 3\cdot14$.)

State why T lies outside the triangle ABC.

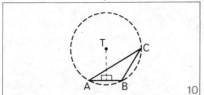

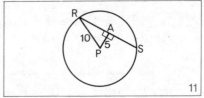

29 In Figure 11, P is the centre of a circle of radius 10 cm. PA is perpendicular to the chord RS and is 5 cm in length.

a Calculate the length of the chord RS to one decimal place.
b State the locus of midpoints of chords equal in length to RS.

30 Draw a circle of radius 4 cm. Use your ruler and protractor to inscribe a regular pentagon in it.

31 PQ is an axis of bilateral symmetry of a circle with centre O and radius 6 cm. A is a point on the circumference of this circle such that AO makes an angle of 45° with the axis of symmetry. Under reflection in PQ, A↔A'. Calculate the length of AA' to two significant figures.

32 Tabulate the order of rotational symmetry and the number of axes of symmetry for the following figures.

a an equilateral triangle   b a square
c a regular hexagon   d a regular octagon.

Can you see any connection between your answer and the number of sides in each figure?

What order of rotational symmetry and how many axes of symmetry would you expect to find in a regular polygon of n sides?

33a Write down the equations of circles with centre the origin and radii 3, 5 and 9.

  b What is the length of the radius of the circle $x^2 + y^2 = 1$?

  c What is the equation of the circle, centre O, passing through the point $(-3, 3)$?

34 Describe the locus of points represented by:

  a $\{(x, y): x^2 + y^2 \geqslant 25\}$ *b* $\{(x, y): x^2 + y^2 < 100\}$

Show on a sketch the intersection of these two sets.

35 If $\triangle PQR$ in Figure 12 is a dilatation of $\triangle ABC$, find the scale factor and the values of p and r.

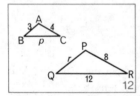 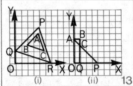 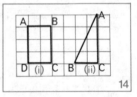

36 In Figure 13(i) $\triangle PQR$ is a dilatation of $\triangle ABC$, and in Figure 13(ii) $\triangle PQC$ is a dilatation of $\triangle ABC$. In each case find the scale factor and the coordinates of the centre of dilatation.

37a Copy Figure 14 (i) on squared paper. Draw the image of rectangle ABCD under a dilatation with centre A and scale factor 2.

  b Copy Figure 14 (ii) and draw the image of triangle ABC under a dilatation with centre B and scale factor $-\frac{1}{2}$.

38 Draw a circle with centre O and radius 4 cm. Draw the image of the circle under a dilatation:

  a $[O, 1\frac{1}{2}]$ *b* $[O, -\frac{1}{2}]$.

39   A, B and C are the points (3, 4), (5, −1) and (−3, 0) respectively. If O is the origin find the images of A, B and C under the dilatations:

a [O, 2]   *b* [O, −3]   *c* [O, 1]   *d* [O, ½].

40a List six points on the line with equation $y = 4x + 1$.
  b List the coordinates of their images under the dilatation [O, 2].
  c Write down the equation of the image of the line $y = 4x + 1$ under this dilatation.

41   A, B, C and D are the points (1, 2), (5, 8), (−1, 12) and (−5, 6) respectively. Find in component form the vectors represented by $\overrightarrow{AB}$, $\overrightarrow{BC}$, $\overrightarrow{DC}$ and $\overrightarrow{AD}$. What can you deduce about the magnitude and direction of these vectors?

  Prove that $\angle\,BAD = 90°$. What kind of figure is ABCD?

42   A is the point (1, 1) and B is (4, 4). Show that the equation of the locus of P(x, y) given by {P: PB = 2PA} is a circle with centre the origin, and state the length of its radius.

Arithmetic

Social Arithmetic 2

1 The value of money

As *members of a family* we all use money to buy food, clothes, entertainment, houses, etc.

As *citizens in a community* all householders have to pay rates to the Town or County Council or Regional Authority to cover part of the expenditure on *local* services such as police, education, street lighting, public libraries and housing.

As *citizens of a country* all wage and salary earners have to pay taxes to the Government to cover expenditure on *national* services such as health, welfare, housing, education, roads and defence.

As a *nation* we need money to pay for national commitments and to carry on trade with other countries in the world.

In this chapter we study some features of these statements. These must be considered, however, against the background of the changing value of money, which is illustrated in Figure 1.

Note on Topics to Explore

It is not possible to study in detail every aspect of social arithmetic, that is, applications of arithmetic in society. Throughout this chapter, therefore, topics of interest and importance are suggested for exploration. Individual pupils, or groups of pupils, might choose one or two of these to investigate, and then report their findings to the rest of the class; in this way everyone can gain a lot of useful ideas and information.

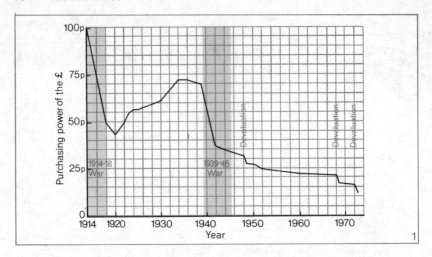

Exercise 1

Figure 1 shows changes in the purchasing power of the £ from 1914 to 1972. Use it to answer the following questions.

1   What is the general 'trend' of the graph?

2   Between which years was the trend upward?

3   What connection do you see between the purchasing power of the £ and the war years 1914–18 and 1939–45?

4   In which years was the purchasing power   *a*   50p   *b*   25p?

5   What was the percentage fall in the value of the £ between 1914 and 1972?

Topic to explore

We often hear about rising wages and inflation. Try to find out the connection between these and the graph.

2 Money and the household

(i) Income and expenditure

Exercise 2

1 Household income consists mainly of the earnings of the wage earners or salary earners in the family. Certain deductions are made from the basic wage or salary; these are shown in the table, which illustrates three typical weekly pay slips.

	GROSS PAY		DEDUCTIONS				NET PAY
	Basic wage	Overtime pay	Income tax	National insurance	Graduated pension	Super-annuation	
A	17.75	4.60	2.60	0.88	0.64	—	18.23
B	45.50	—	5.65	1.00	1.35	2.73	34.77
C	38.00	11.40	2.82	0.88	1.85	—	43.85

a Check that in each case the net pay has been calculated correctly, by subtracting the deductions from the gross pay.

b Calculate B's annual wage (52 equal weekly payments).

c How much income tax is paid annually by C?

2 A salesman has a fixed annual salary of £1200, plus a commission of $2\frac{1}{2}\%$ on all his sales. If his total sales in one year amounted to £32 000, calculate his annual income before deductions of tax, etc.

3 A shopkeeper estimates that after paying all expenses his profit is 7% of his takings. Calculate his estimated weekly profit if the year's takings were £39 000.

4 Figure 2 illustrates the changes in weekly wage rates and certain prices during the years 1969, 1970 and 1971. These measures are all based on a figure of 100 in 1963.

a What is the general trend of all these measures?

b Which price has risen least during the period 1969–71?

c Compare the graphs of weekly wage rates and retail prices; comment.

d Can you see a seasonal trend in the increases in weekly wage rates each year?

e What are the percentage increases in weekly wage rates and wholesale prices shown over the period 1963 (when each measure was 100) to the end of 1971?

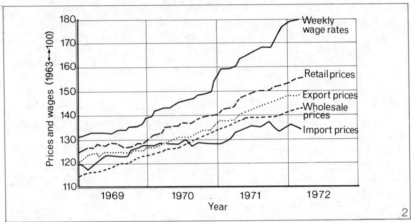

5 A household's average weekly budgets in 1965 and 1972 are shown below.

Expenditure	1965	1972	Net income	1965	1972
Rent and rates	£2·65	£ 3·80	Husband	£19·10	£26·60
Food	9·75	12·00	Wife	9·60	13·40
Clothes	3·80	5·00			
Transport	1·45	1·95			
Insurance	0·65	0·65			
Entertainment	2·77	3·10			
Miscellaneous	3·53	5·25			

a How much did the family save weekly in 1965 and in 1972?

b What percentage of the total net income was spent on food in 1972?

6 The index of retail prices for food and housing at the end of each year from 1962 to 1971 is given below. These are based on a measure of 100 in January 1962; 102 indicates a 2% increase during 1962, and so on.

Year	1962	1963	1964	1965	1966	1967	1968	1969	1970	1971
Food	102	105	108	112	116	119	123	131	140	160
Housing	102	104	107	109	113	118	141	147	158	175

a In which year did the housing index overtake the food index?

b In which years did each increase most?

c What is the percentage increase of each from January 1962 (value 100) to the end of 1971?

d Calculate the percentage increase in the food index during 1971.

Topic to explore

Investigate the payments for, and the benefits that can be claimed from, National Insurance and the Graduated Pension Scheme.

(ii) Holiday travel

Each country has its own currency. In Britain, for example, the unit is £1 sterling, and in the United States of America it is the dollar. The financial columns of the newspapers show the *rates of exchange* between various currencies; these rates change from time to time. Here are a few rates of exchange, as at 20th October, 1972.

Country	Currency	Number per £1
America	Dollar	2·39
Austria	Schilling	55·00
Belgium	Franc	104·50
Denmark	Krone	16·46
France	Franc	11·70
Germany	Deutschmark	7·62
Italy	Lira	1425·00
Spain	Peseta	150·50
Switzerland	Franc	9·00

When travelling from one country to another you 'buy' the currency of the country to which you are going. To do this, the money involved has to be converted from one currency to another by means of the appropriate rate of exchange. The calculation is an application of *direct proportion*.

Example. A tourist in France changed £75 into francs at the rate of 11·70 francs to the £. He spent 710·50 francs and changed the remainder into sterling at the rate of 11·90 francs to the £. How much sterling did he receive?

(i)
$$
\begin{array}{lcl}
\text{Number} & & \text{Number} \\
\text{of £} & & \text{of francs} \\
1 & \longleftrightarrow & 11\cdot70 \\
75 & \longleftrightarrow & 11\cdot70 \times \dfrac{75}{1} \\
& = & 877\cdot50 \text{ francs}
\end{array}
$$

He spent   710·50 francs
Remainder   167·00 francs

(ii)
$$
\begin{array}{lcl}
\text{Number} & & \text{Number} \\
\text{of francs} & & \text{of £} \\
11\cdot90 & \longleftrightarrow & 1 \\
167 & \longleftrightarrow & 1 \times \dfrac{167}{11\cdot90} \\
& = & £14\cdot03
\end{array}
$$

He received   £14·03

Note. It must be realised that use of logarithms or a slide rule is not usually appropriate in money questions, where 3 significant figures may not be sufficient. However, if you set up the calculations for the example above on your slide rule, you will realise that in fact you can use the slide rule as a ready reckoner in Foreign Exchange questions when approximate answers are sufficient.

Topics to explore

1 *Look up the rates of exchange in a newspaper and make a list of countries, their currencies and rates of exchange. Compare your list with the list in the paper at a later date.*

2 *From your atlas choose a holiday resort abroad. Make a record of the route you would take to get there, the approximate distances by land, sea or air, and if possible the travelling times. Decide the amount of money you would take to spend, convert it into the appropriate currency, and make up a record of spending.*

3 *Obtain a 'holiday travel brochure', plan a holiday from it, and work out the travelling and living costs associated with it.*
   (These projects may replace some of the examples in Exercise 3.)

Exercise 3

In the following examples select the rate of exchange from the table given at the beginning of the Section, where necessary.

1 How many dollars would be obtained for £30 sterling?

2 How many Danish kroner would be obtained for £50?

3 A watch costs 35 francs in Switzerland. How much is this in British currency?

4 The pupils in a school party visiting Rome were each allowed to take £12 spending money. How many lire is this?
   One girl spent 14 100 lire. How much British money will she receive for the lire she has left?

5 A tourist took £100 to Germany. He changed this to Deutschmark at the rate of 7·60 DM to the £, spent 606 DM, and then changed

the remainder to British money at the rate of 7·70 DM to the £. How much of his £100 was left?

6 Here is an extract from a typical tourist company leaflet. It shows the inclusive charges per person for a 14-day holiday:

Places and hotels	From Glasgow	From London
Paris: Hotel A	£70	£65·50
Hotel B	62	57·80
Rome: Hotel A	93·20	87
Hotel B	80	77·60

Calculate the cost of a fortnight's holiday for:

a one person in Hotel A in Paris, flying from Glasgow
b two people in Hotel B in Rome, flying from London
c three people in Hotel B in Paris, flying from Glasgow at a peak period when an extra daily charge of £0·50 per person per day is made.

7 Work out the following hotel bill in Cologne, expressing the total sum to be paid in British currency.

12 days room at 10 Deutschmark per day
24 breakfasts at 3 DM each
10 lunches at 8·50 DM each
24 dinners at 12 DM each
A 10% service charge is added to the bill.

8 A man went to the Continent for a holiday. He started with £100 which he changed into French currency. He spent 300 francs before going on to Spain where the rate of exchange was 1 franc to 12 pesetas. In Spain he spent 9440 pesetas. How much British money did he arrive home with?

9 Two friends went on a Coach Tour to Interlaken, Switzerland, costing £33·50 per person. During their holiday they spent 400 French francs and 900 Swiss francs altogether. Find the total cost of the holiday (to the nearest £).

10 Calculate the cost of a holiday for two at Stresa, Italy:

Car air ferry Southend to Geneva:
car £15 single, driver and passenger £33·45 each return.
21 days room at £3·30 per day
42 breakfasts at £0·40 each
20 lunches at £0·85 each
42 dinners at £1·25 each

Stresa is 250 km from Geneva. Petrol consumption is 8 litres per 100 km, and petrol costs 8p per litre in Switzerland and in Italy.

(iii) Hire purchase and credit sales

Many people buy goods on credit, perhaps a radio or a bicycle, perhaps something more expensive like furniture or a motor-car. Credit enables them to have the goods at once and to pay for them out of their earnings later.

There are two main kinds of arrangement—hire purchase and credit sales; in both, payment is made by a deposit and instalments. Under hire purchase the goods do not become the buyer's property until all the instalments have been paid; until then they belong to the shopkeeper or finance company. Under a credit sales agreement the goods become the buyer's property when the deposit is paid.

Example. Figure 3 shows the kind of advertisement you see in newspapers and magazines. Calculate the extra cost by hire purchase.

Only £3·90 down &
12 monthly
instalments of
£0·94

CASH PRICE £13·95

3

Cash price of radio	= £13·95
By hire purchase, deposit	= £3·90
Cost of instalments	= 12 × £0·94
	= £11·28
Total cost	= £3·90 + £11·28
	= £15·18
Extra cost by H.P.	= £15·18 − £13·95
	= £1·23

Topic to explore

Cut out some hire purchase advertisements and study the various terms offered; perhaps calculations could be based on these instead of on the examples in the following exercise.

Exercise 4

1   A camera can be bought for £25 cash, or by paying 12 monthly instalments of £2·30. Calculate how much extra is paid in the latter case.

2   A washing machine costs £89·45 cash, or a deposit of £25 and 24 monthly payments of £3. How much extra is paid in the latter case?

3   A summer-house can be bought for £36·70 cash, or by a deposit of £5·52 plus 24 monthly payments of £1·60. What is the difference in cost between the two methods of paying for the article?

4   A man buys a car by putting down a deposit of £175 and paying 117 weekly payments of £3. If the cash price of the car is £494, find how much less the man would have paid by paying cash for the car.

5   An Air Company advertises special holiday rates on their flights to Canada at £83·50 cash or a deposit of £9·50 plus 24 monthly payments of £3·50. Find the difference in cost.

6   A TV set costs £100 cash. The set can be paid for by instalments if a deposit of 25% of the cash price is paid. How much would the equal monthly payments be if the set is paid for in 1 year?

7   A 14-day air charter holiday in the U.S.A. and Canada costs £198·45 cash, or a deposit of £38·45 and 12 monthly repayments of £14·67. Find the difference in cost.

8   A gas cooker is advertised at £55·35 cash or £68·62 by hire purchase. For hire purchase there is a deposit of £5·62, and 15 quarterly payments. Find the repayment due each quarter.

9   A colour television set can be purchased for £350 cash, or by a down payment of 20% of the cash price plus 25 monthly instalments of £13·50. Find the difference in cost between the two methods, and express the difference as a percentage of the cash price (to the nearest 1%).

3 Safeguarding money

(i) Savings and investment

Suppose that in 1946 a man had £1000 to invest for a period of 25 years. Here are several ways in which he might have invested it.

Investment	Value of his £1000 in 1971
(i) Buried it in the garden	£1000
(ii) Banked it at $2\frac{1}{2}\%$ compound interest	£1853
(iii) Building Society at 5% compound interest	£3387
(iv) Bought a house	£5100
(v) Bought stocks and shares	£6300

Now look back at Figure 1. What happened to the purchasing power of £1 between 1946 and 1971? Note that the original £1000 is worth only £380 in 1971.

Figure 4 shows the approximate values of the investments over the 25-year period. Which methods of investment were worth while?

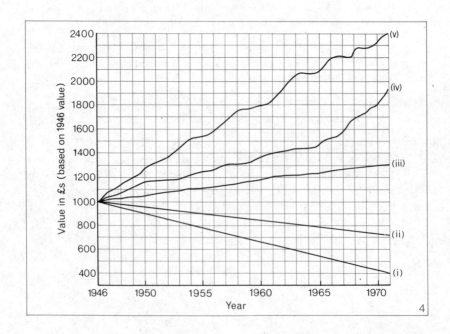

Personal savings can be invested in a variety of ways which have various advantages and disadvantages. Some forms of investment safeguard your money for a fixed number of years, and others allow money to be withdrawn after a certain period if notice has been given. In some, both the value of the investment and the interest on it vary with the fluctuations of the market.

Some ways of saving and investing money

1 Trustee Savings Bank
In the Ordinary Department the investment can be withdrawn on demand; the rate of interest is 3–4% and the first £21 of interest is free of income tax. In the Special Investment Department notice is required for the withdrawal of large sums of money, and the rate of interest is 7–9%.

2 Deposit Account in a Bank
No notice is required for the withdrawal of cash from deposit accounts, but notice is usually required in the case of special investment accounts. The rate of interest is 7–10%.

3 Savings Account with a Building Society
Investors can usually withdraw small sums on demand, and larger sums after giving notice. The interest rate is 5–7%, and income tax on this is paid by the Society.

4 National Savings Certificates
The Decimal Issue of certificates cost £1 each, and are worth £1·25 each at the end of four years. They can be cashed at any time, but their growth is planned to discourage this before several years have elapsed. The rate of interest over the full period is 5·7%.

5 Premium Savings Bonds
No interest is paid on these bonds, which cost £1 each, but the holders participate in a weekly 'draw' for prizes of up to £25 000; in addition, there is a monthly prize of £50 000.

6 Save-As-You-Earn
A wide variety of S.A.Y.E. schemes exist whereby a fixed sum of money is invested regularly for five or seven years, and bonus payments are made at the end of these periods, providing rates of interest of 7·2% and 7·6% respectively. If the scheme is linked with life assurance the investor can obtain a rebate of income tax.

7 *Stocks and Shares*
Government stocks, or shares in companies, can be bought through
a stock broker, or sometimes through a bank. Both the value of the
investment and the amount of interest, or dividend, may fluctuate
with the fortunes of the company concerned. A person selling shares
at a profit may be liable to pay capital gains tax.

8 *Unit Trusts*
In buying units in these trusts an investor is employing the expertise
of the trust managers to invest his money in a wide variety of stocks
and shares. The value of the units and the dividend can fluctuate,
and again there may be liability to pay capital gains tax on selling.

9 *Life Assurance*
A life assurance with-profits policy not only gives insurance cover
but also provides a cash bonus when the policy matures. The
premiums, which are usually paid quarterly or annually, can qualify
for income tax relief.

Illustrations of most of the above are included in Exercise 5 and
later Exercises in this chapter.

Topics to explore

*Savings and careful investments are always worth while. Find out more
about savings and investments, e.g. National Savings Certificates,
Unit Trusts, S.A.Y.E.*

*Select a share or Unit Trust quoted in the financial column of a
newspaper and graph its market value daily or weekly over a period of
time.*

Exercise 5

1 A Savings Bank offers $7\frac{1}{2}\%$ per annum interest in their Special Investment Department. Find the interest on £240 for 10 months.

2 A Building Society offers 5% per annum interest in its Shares Investment Account. Calculate the interest, to the nearest £, on £500 for
200 days.

3 A Savings Bank offers an interest rate of 4% per annum, the first £21
of interest being free of income tax.

a What interest would be paid on £250 for 1 year?

b What sum of money would yield £21 interest in 1 year?

4 A recent issue of National Savings Certificates offers the following. Purchase price per unit, £1. Each unit gains:

£0·025 at the end of the first year

£0·012 for each complete 4 months during the second year

£0·021 for each complete 4 months during the third, fourth and fifth years.

How much are 1000 units worth at the end of 1, 2, 3, 4 and 5 years?

5 The market value of a Unit Trust share was quoted in the daily press at 40p, and the half-yearly dividend on each share was 0·828p.

a What is the dividend paid to a person with 100 shares?

b Express the ratio of the dividend per share to the market value of a share as a percentage.

6 A Government offered £3 million new stock, bearing interest of $6\frac{1}{2}\%$ per annum. How much interest must the Government pay out each year?

7 Shares in a certain company cost £3 each and pay an annual dividend of £0·15 per share.

a What percentage return is this on the investment?

b How many (whole) shares can be bought with £100?

c What annual income is then obtained?

8 In a Save-As-You-Earn scheme a man agrees to make 60 monthly payments of £10. At the end of five years he will receive his money back, along with a bonus to the value of 12 monthly payments; if he leaves his money for two more years this bonus will be 24 monthly payments instead.

a How much would he receive at the end of 5 years, and 7 years?

b To what average rate of interest per annum over the 7 years is the last two years' bonus equivalent?

9 A man buys a house and has to borrow £1500 from a Building Society. He pays back £185 at the end of one year; this covers the interest charge on the loan of 7%, and also reduces the loan. How much does he owe the Society at the end of the year?

10 A man borrowed £2000 from a Building Society. The rate of interest was 8%, calculated on the amount outstanding at the beginning of a year. He repaid £535 during the first year, which covered the interest charge and reduced the loan. How much did he owe at the beginning of the second year?

How much would he have to repay at the end of the second year in order to reduce the loan to £1000?

11 A Trustee Savings Bank pays interest of $3\frac{1}{2}\%$ per annum on Ordinary accounts and $6\frac{1}{2}\%$ per annum on Special accounts, both paid on complete £s in the account. How much interest is due to a person who had £72·16 in the Ordinary account and £340·67 in the Special account for a year?

12 In 1971 the National Debt was £33 440 million. How much interest at 6% per annum would have to be paid on this debt?

13 In the Decimal Issue of National Savings Certificates each £1 unit grows as follows:
At the end of the first year by 3p to £1·03.
During the second year by $1\frac{1}{2}$p for each 4 months to £1·07$\frac{1}{2}$.
During the third year by $2\frac{1}{2}$p for each 4 months to £1·15.
During the fourth year by 3p for each 4 months, and 1p at the end to £1·25.

Calculate the rate of interest paid each year, to 2 significant figures, based on the value of a unit at the beginning of that year.

(ii) *Compound interest*

When money is deposited in a bank for a period greater than one year it is customary for the interest to be added at the end of every year. For example if a principal of £100 is deposited, and gains interest at the rate of 5% per annum, then at the end of one year the interest is £5. This is added to the original principal so that the principal for the second year is £105, and the interest for the second year will be 5% of £105, that is £5·25. Now the principal for the third year will be £110·25 and so on.

This is an example of *compound interest*. Notice that, in compound interest, the principal increases every year, so that the interest also increases every year. As with simple interest, compound interest is normally calculated on complete pounds of principal.

Example. Calculate the compound interest on £450 for 3 years at 4% per annum.

First principal	=	£450	1% of £450 = £4·50
First interest	=	18	4% of £450 = £18·00
Second principal	=	468	4·68
Second interest	=	18·72	18·72
Third principal	=	486·72	4·86
Third interest	=	19·44	19·44
Amount	=	506·16	
Original sum	=	450	
Compound interest	=	£56·16	

Note. Compound growth and depreciation occur in many situations. For example, the population of a country may increase by 2% each year, or the value of machinery may decrease by 25% each year. Calculations are similar to the one shown above.

Exercise 6

Calculate the compound interest on

1   £500 for 2 years at 4% p.a.   *2*   £350 for 2 years at 2% p.a.

3   £540 for 2 years at 5% p.a.   *4*   £800 for 3 years at 2% p.a.

5   £642 for 3 years at 6% p.a.   *6*   £483 for 4 years at 10% p.a.

7   Calculate the compound interest on a deposit of £640 for 3 years if the rate per cent per annum is 4% for the first year, 5% for the second year and 6% for the third year.

8   £1000 is invested for 2 years at 4% per annum. Calculate:

a   the simple interest, i.e. calculated on the principal of £1000 each year
b   the compound interest, i.e. calculated on 'principal + interest' at the beginning of each year.
  What is the amount at the end of 2 years in each case?

9   The population of Britain at the end of 1972 was estimated to be 56 million. If the growth rate is 3% per annum, what is the estimated population for the end of 1973 and 1974? (Work to 4, and round off to 3, significant figures.)

10 A car cost £1200 when new. Calculate its value, to the nearest £, after 3 years, assuming an annual depreciation of 25%.

11 Repeat question *10* for a car worth £1000 when new, and an annual rate of depreciation of 20%.

12 A man borrowed £800 at a rate of interest of 5% per annum, which he had to pay each year on the sum of money outstanding at the beginning of the year.

At the end of the first year and again at the end of the second year he paid back £300, part of which paid the interest, the rest of which reduced the loan. How much had he to pay at the end of the third year to clear the debt and interest due?

13 A formula for calculating the amount £*A* of a principal £*P* after *n* years at a rate of *r*% per annum compound interest is

$$A = P\left(1 + \frac{r}{100}\right)^n.$$

Use this formula, and logarithms, to calculate the answers to questions *1–6*.

(*iii*) *Insurance*

Insurance is a contract whereby a company agrees to pay compensation for injuries, fire, theft, etc., in return for an agreed sum of money called the *premium*. In general, insurance concerns events which may or may not happen; for example, one can insure a car or house against fire, accident, etc.

Assurance is used in connection with payments and policies concerning events which are certain to happen. There are in fact close connections between the ideas and methods of probability and the calculations of premiums in assurance. An assurance policy is usually taken out on a person's life. There are two main types: an 'endowment' policy for a fixed period, and a 'whole life' policy. In the former the assured person will receive a fixed sum of money if he survives the term of the endowment, and his dependants will receive it if he does not survive. In the case of a 'whole life' policy, the policy is payable on death; so the sum assured is certain to be paid sometime.

(a) Life assurance

ENDOWMENT ASSURANCE

Age next birthday	Annual payments to assure £100 (with profits) at the end of		
	20 years	25 years	30 years
20	£5·35	£4·33	£3·63
21	5·35	4·33	3·63
22	5·37	4·35	3·65
23	5·37	4·35	3·65
24	5·40	4·38	3·67
..			
30	5·45	4·43	3·70
40	5·75	4·75	4·02
50	6·33	—	—

This table shows the annual premiums payable for an endowment assurance policy worth £100. Policies with or without profits can be obtained and, whereas the premiums for the former are slightly higher, the profits payable help to overcome the falling value of the £ shown in Figure 1. The company makes a profit through its investments and through premium policy payments, and passes a proportion on to the 'with profits' policy holders.

Example. A man aged 21 wishes to take out an endowment assurance policy (with profits) for £2500, payable in 20 years. Calculate the annual premium.

$$\begin{array}{ccc} \textit{Sum assured} & & \textit{Premium} \\ £100 & \longleftrightarrow & £5·35 \\ £2500 & \longleftrightarrow & £5·35 \times \dfrac{2500}{100} \\ & = & £133·75 \end{array}$$

(In practice the amount he would effectively pay would be about £113 owing to income tax relief on the premium. Thus over the 20 years the man will pay a total of approximately £20 × 113, i.e. £2260; and since he had £2500 assurance cover on his life during the 20 years, he will receive the £2500 plus profits at the end of the period.)

The following table shows premiums payable to assure £100 on death.

WHOLE LIFE ASSURANCE

Age next birthday	With profits	Without profits
20	£2·05	£1·42
21	2·12	1·45
22	2·20	1·50
23	2·25	1·54
24	2·30	1·60
..		
30	2·63	1·75
40	3·41	2·61
50	4·75	3·77

Topics to explore

1 An up-to-date method of saving is to take out Life Assurance through a Unit Trust. Here the citizen undertakes to make a regular payment, part of which is used to assure against death, and the rest is used to purchase 'Units'. Find out the details of such a scheme.

2 Most people borrow money to buy houses. Find out details of some of the ways of borrowing this money, e.g. Building Society mortgages, Local Authority loans, Insurance Endowment mortgages, etc.

Exercise 7

1 From the tables above give the premiums payable to assure £100 for:

Type of policy	Age next birthday	With/without profits	Number of payments
a Endowment	40	With	25
b Endowment	23	With	30
c Whole life	22	With	—
d Whole life	24	Without	—

2 A man aged 23 next birthday wishes to take a whole life policy for £4500 without profits. Calculate his annual premium.

3 Find the annual premium for an endowment policy of £2200, with profits, payments being made over a 30-year period by a person aged 21 next birthday.

4 A man aged 22 next birthday takes out a whole life policy for £1800 with profits. Calculate his annual premium.

5 A man aged 30 next birthday takes out a whole life policy, without profits, of £1000. If he dies after making 12 payments, find the difference between the amount of money payable and the amount he had paid in premiums.

6 Figure 5 shows the premiums payable on a whole life assurance of £100 (with and without profits) by persons starting payments at ages from 20 to 50. What general information is given by the graphs?

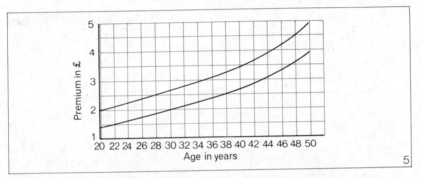

7 A man aged 40 on his next birthday calculates that he can invest £35 a year for the next 20 years in a with-profits endowment policy. What is the greatest amount for which he can assure his life?

8 If instead the man in question 7 took out a whole life with-profits policy, for what sum could he assure his life? For how much more could he assure his life if he chose a without-profits policy?

(b) Property insurance

Example. A householder wishes to insure his house and its contents. The house is valued at £4800 and the contents at £2250. Calculate the annual premium payable if the rate is £0·12 per cent for buildings and £0·25 per cent for contents.

	Sum insured		*Premium*
House	£100	$\longleftrightarrow$	£0·12
	£4800	$\longleftrightarrow$	£0·12 × $\dfrac{4800}{100}$
		=	£5·76
Contents	£100	$\longleftrightarrow$	£0·25
	£2250	$\longleftrightarrow$	£0·25 × $\dfrac{2250}{100}$
		=	£5·63
Total annual premium		=	£11·39

Exercise 8

1   Calculate the premium on:

a   £2500 at £0·12 per £100   *b*   £3750 at £0·50 per £100

c   £750 at £0·20 per £100   *d*   £5650 at £0·22 per £100

2   A man wishes to insure his house and contents. The house is valued at £7500 and the contents at £3250; find the yearly premium if the rate is £0·12 per cent for buildings and £0·20 per cent for contents.

3   I insure my watch for a premium of $2\frac{1}{2}\%$ of its value. If my watch is valued at £50, what premium do I pay annually?

4   What is the annual premium to insure a house worth £9500:

a   against fire only at £0·15 per cent
b   against all risks at £0·22 per cent?

5   Find the total premium I have to pay to insure a caravan worth £550 and its contents worth £150 at the following rates: £1·30 per £100 for the caravan, and £1 per £100 for the contents.

6   A householder has his house valued at £6500 and its contents at £2500. Calculate the total annual insurance premium at the rate of £0·12 per cent for the house, and £0·25 per cent for the contents, if he is allowed a 10% discount on the total.

7   An insurance company charges the following rates (per £100) for property insurance:

> *Buildings*: comprehensive—15p, fire only—$7\frac{1}{2}$p
> *Contents*: comprehensive—25p, fire only—12p
> *Valuables*: jewellery, etc.—£1·25

Find the total premium paid by a man who insures his house (worth £8800) against fire only, the contents (worth £3200) and his wife's jewellery (worth £260) against all risks.

8   Using the rates given in question 7 find the premium for each of the following:

a   House—value £9200; comprehensive insurance.
b   Furniture—value £1400; fire only insurance.
c   Amplifier and loudspeakers—value £140; comprehensive insurance.
d   Business premises—value £150 000; fire only insurance.

(c) Motor vehicle insurance

Each insurance company has its own basic rates for motor vehicle insurance, with its own set of discounts or surcharges which are applied to the basic rates. The basic premium for motor-cars depends on two main factors:
(i) the make and model of car
(ii) the part of the country where the car is kept and used.

The tables show the basic premiums for three classes of car in low, medium and high-rated areas of the country, and the rates of 'no-claims bonus', which is a discount given after a stated number of years in which no accident claim has been made.

Area	Classification of car			Rate of no-claims bonus	per cent
	2	4	6		
Low rated	£38	£55	£70	After 1 year	$33\frac{1}{3}$
Medium rated	£42	£64	£81	After 2 years	40
High rated	£50	£72	£94	After 3 years	50
				After 4 years or more	60

Exercise 9

1 Find the motor-car insurance premiums paid by owners of cars as follows:

	Area	Classification	Number of years without a claim
a	Low rated	4	3
b	High rated	2	2
c	Medium rated	6	1

2 A young man has just bought his first car which has a 2 classification. He lives in a medium-rated area, and because of his inexperience the insurance company adds a 50% surcharge to his premium. How much has he to pay?

3 A man owns a class 6 car in a low-rated area. He has had 5 years without a claim and is entitled to a further 10% discount on the premium *after* deduction of the no-claims bonus. How much has he to pay?

Topic to explore

Find out about car and motor-cycle insurance-policy coverage. Companies which offer low premiums may not give as good cover or may have more stringent conditions applying to claims. Grounds on which a company might restrict its cover, charge more, or not accept insurance are: a person's age, occupation, accident proneness and inexperience.

---

4 Money and the Council

(i) Local rates

Here is a list of estimated expenditure in Edinburgh in the year 1972–73:

Services	Expenditure (£)	Services	Expenditure (£)
Police	2 424 594	Education	20 540 685
Lighting	916 347	Welfare	4 003 345
Cleansing	2 180 142	Health	1 176 940
Public baths	426 276	Housing	1 400 358
Public laundries	107 104	Planning	482 114
Public libraries and museums	630 631	Civil defence	3 201
Public parks	1 087 419	Miscellaneous services (fire; voters' register;	
Burial grounds	76 596	births, marriages, deaths	
Streets and footpaths	3 550 354	register, etc.)	3 277 327
Sewers and drains	690 374	Contingent expenditure	1 924 950

Amount of estimated expenditure in respect of 1972–73	44 898 757
Less Estimated Government Grants	16 550 030
Total amount to be raised in respect of City Rates	28 348 727

City Rates	per £
City Rate (including public water rate)	0·73
Domestic water rate	0·04
Total	£0·77

At the foot is the *City Rate* of £0·77, or 77p in the £. Rates are a local tax levied on the occupiers of houses, shops, business premises and offices in order to meet the above expenditure. The houses, shops, etc., are assessed at a certain *rateable value*, depending on the annual rent at which the property could be let, and each £ of rateable value of property in Edinburgh was charged at the rate of 77p for the year 1972–73.

Example. Calculate the rates payable by a householder in Edinburgh if the rateable value of his house is £128.

$$
\begin{array}{lcl}
\textit{Rateable value} & & \textit{Rates} \\
£1 & \longleftarrow & £0{\cdot}77 \\
£128 & \longleftarrow & £0{\cdot}77 \times \dfrac{128}{1} \\
& & = £98{\cdot}56
\end{array}
$$

Topic to explore

Local expenditure on rates is closely related to social factors. Find out how the money is spent in your town or county or region, what the rate per £ is, and how the rate has changed over the years.

Exercise 10

1   Calculate the rates payable by a householder in Edinburgh whose house has a rateable value of £80. (Rate 77p in the £.)

2   Calculate the rates payable in each of the following:

	a	*b*	*c*	*d*
Rateable value	£130	£85	£140	£180
Rate per £	80p	70p	93p	66p

3   The rate in one region was raised from 65p to 77p in the £. How much more had to be paid on a house with rateable value £135?

4   In 1966 the rateable value of a house was £40, and the rate was £1·25 per £. After revaluation, the rateable value of the house was £80, and the rate was £0·82 per £. How much more or less had the householder to pay now?

5   A family moved from a house with a rateable value of £115 in an area with a rate of 80p per £ to a house with a rateable value of £220 in an area with a rate of 60p per £. How much more had they to pay in rates?

6   What percentage of the estimated expenditure in Edinburgh was due to expenditure on education? (Take both costs to the nearest £100 000.)

7   The rent of a Council house was £55 per annum, and rates had to be paid at £0·99 per £ on a rateable value of £80. Find to the nearest necessary £0·01 the weekly payments (50 only are made in a year) to the Council to cover rent and rates.

8   Repeat question 7 for a rent of £76 per annum, and a rateable value of £108.

(ii) Calculating the rate per £

Example. The total rateable value of a town is £1 750 000, and the total expenditure is estimated to be £1 950 000. Calculate what rate per £ must be charged.

Rateable value, in £		Rates, in £
1 750 000	$\longleftarrow$	1 950 000
1	$\longleftarrow$	$1\,950\,000 \times \dfrac{1}{1\,750\,000}$

$$= \frac{39}{35}$$

$$= 1{\cdot}114...$$

The rate charged must be £1·12 per £, as £1·11 would not provide a sufficient income to cover the expenditure.

Exercise 11

1   Calculate the rate per £ in each of the following:

	Total rateable value of area	Estimated expenditure
a	£750 000	£600 000
b	£1 000 000	£1 050 000
c	£1 250 000	£950 000

2   For each case in question *1*, how much would a householder have to pay on a rateable value of £100?

3 a   What rate per £ must be levied to raise £455 000 from a total rateable value of £600 000?

  b   What rates would a householder pay on a house with rateable value £118?

4   A region has a rateable value of £23·5 million, and an estimated expenditure of £18·5 million. Calculate the rate per £ to be charged.

5   A city has a rateable value of £2·2 million, and an estimated expenditure of £1·9 million. Calculate the rate per £ that must be charged, and the surplus over £1·9 million that this rate will raise.

Topics to explore

1   *The whole matter of Government income and expenditure is as much a social as a mathematical matter. Study the reasons for spending and taxing in the way these are done; the rates of certain taxes, and even the taxes themselves, change from time to time. It would also be of interest to find out how other countries balance their national budgets.*

2   *Another point: some people think that expenditure in local Town and County Council rates on items like education and housing should be met entirely from Government funds; this would probably mean that the money would come from the taxpayer instead of from the ratepayer. Does this make any difference?*

5 Money and the Government

(i) National income and expenditure

Just as each family has to pay its accounts and to try not to spend
more money than it has, and just as citizens have to pay rates to the

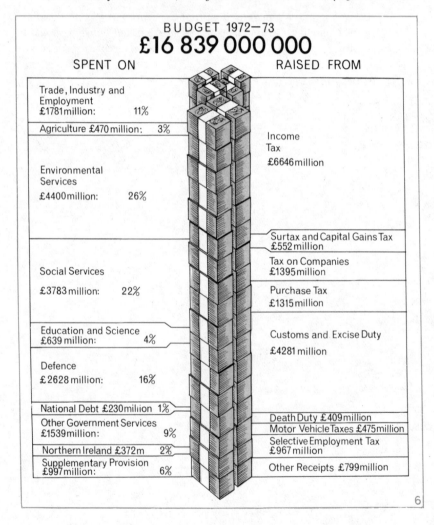

BUDGET 1972—73

£16 839 000 000

SPENT ON   RAISED FROM

Trade, Industry and
Employment
£1781 million: 11%

Agriculture £470 million: 3%

Income
Tax
£6646 million

Environmental
Services
£4400 million: 26%

Surtax and Capital Gains Tax
£552 million

Tax on Companies
£1395 million

Social Services
£3783 million: 22%

Purchase Tax
£1315 million

Education and Science
£639 million: 4%

Customs and Excise Duty
£4281 million

Defence
£2628 million: 16%

National Debt £230 million 1%

Other Government Services
£1539 million: 9%

Death Duty £409 million

Motor Vehicle Taxes £475 million

Selective Employment Tax
£967 million

Northern Ireland £372m   2%

Supplementary Provision
£997 million: 6%

Other Receipts £799 million

6

Regional Authority to pay for local services, so the Government has to pay for national expenditure and to try to balance the national budget.

Figure 6 shows the national income and expenditure for 1972–73.

Early in April each year the Chancellor of the Exchequer presents the budget, on behalf of the Government, to Parliament, and makes a statement of a social, economic and financial nature.

Notice that most of the money is raised from:

(i) taxes on persons and companies—direct taxes; and

(ii) taxes on spending—indirect taxes.

Exercise 12

1   List, in order, the four main items of expenditure and the four main items of income in the 1972–73 budget (Figure 6).

2   What percentage of the total expenditure was for education and science?

3   What percentage of the total income came from income tax?

4   The annual motor-vehicle licence costs £25. What revenue does this provide in a year in which $2\frac{1}{2}$ million vehicles are licensed.

5   Illustrate the various items of budget expenditure for 1972–73 in a pie chart; the percentages are given in Figure 6.

(ii) Income tax

Income tax is levied on all income, i.e. income from wages or salary, from investments, from interest, etc. Income tax is not paid on the whole income, but on the *taxable income*, that is, the income after deduction of certain *allowances*.

Here are some of the allowances for 1973–74:

Lower personal allowance (single person)	£595
Higher personal allowance (married man)	775
Wife's earned-income allowance	595
Children: not over 11 years old	200
11 but not over 16 years	235
16 years or over (in full-time education)	265

There are other allowances, e.g. for life assurance premiums, a housekeeper, a blind person, etc. For example, the life assurance allowance is:

the amount of the premiums if they are less than £10; £10 if the total premiums are between £10 and £20; one half of the premiums if they are more than £20.

Note that the above allowances may suffer deductions to ensure payment of tax on certain other sources of income, including any weekly Family Allowances received.

The rates of tax for 1973–74 are:

30% on the first £5000 of taxable income
40% on the next £1000 of taxable income
45% on the next £1000 of taxable income
50% on the next £1000 of taxable income
55% on the next £2000 of taxable income

and so on to a maximum of 75%.

Example. A married man has a salary of £3000 per annum. He has three children, aged 7, 12 and 17, who are all at school. He pays a life assurance premium of £80 annually. Calculate the monthly income tax deduction from his salary.

Allowances	Higher personal	£775
	7-year-old child	200
	12-year-old child	235
	17-year-old child	265
	Life assurance	40
	Total allowances	£1515

Taxable income		£3000
	less	£1515
		£1485

Income tax = 30% of £1485   = £445·50

Monthly deduction of tax   = £37·12

Exercise 13

In this Exercise use the *allowances* and *rates of tax* given on pages 189–190.

1   Calculate the information required in the last column of this table:

	Income	Married or single	Children's ages	Life assurance premium	Calculate
a	£1610	single	—	£30	Annual tax payable
b	£2500	married	5, 10	£90	Annual tax payable
c	£1980	married	17, 18 (at school)	£80	Annual tax payable
d	£4100	married	10, 12	£180	Monthly tax payable

2   A married man with a salary of £3250 has a daughter aged 7. If his life assurance premiums total £90, how much tax does he pay monthly?

3   A single woman has a salary of £1500. If she receives a 'dependent relative' allowance (for her mother, whom she supports) of £75 per annum, calculate her net annual income (i.e. after deduction of tax).

4   A man borrows £1500 from a Building Society at 7% per annum rate of interest.

a   How much interest is he due to pay at the end of one year?
b   If he is given income tax relief at the rate of 30% of the interest, how much does he actually have to pay?

5   A man takes out a life assurance policy for £2500 at an annual premium of £1·52 per £100. If he is allowed income tax relief at the rate of 30% on half of his total premium, calculate the real annual cost of the insurance.

6   A married man with a son aged 12 and a daughter aged 14 earns £40 per week. Calculate his weekly pay after deduction of income tax.

7   Repeat question *6* for a married man with children aged 1, 4 and 6, and a weekly wage of £50.

8   A married man with children aged 12, 14 and 18 (at University) has a salary of £9000; in addition, he has other income totalling £800 per annum. If he pays £300 annually for life assurance and has additional allowances of £140 each year, calculate:

a   the amount of income tax he pays annually (note rates over £5000)
b   his net monthly income.

(iii) Value added tax (VAT)

Value added tax is a tax which is chargeable on the supply of goods and services in the course of a business and on all imports of goods, except when legislation provides otherwise.

A zero rate is stipulated (i.e. no tax will be charged) for certain goods and services. Examples are food, fuel, books and newspapers, passenger transport, etc.

In this Section we assume that a rate of 10% is charged on all other goods and services. Cars are subject to an additional tax as well as VAT.

In all problems involving VAT it is important to keep the prices *before* and *after* tax separate, as shown in the worked examples.

Example 1. A party of six attend a dinner-dance costing £2·15 per person, before the addition of VAT. How much tax had to be remitted to the Customs and Excise, and what was the total bill paid by the party?

The cost of 6 dinners $= 6 \times £2·15 = £12·90$ (before VAT).

Tax exclusive price	VAT	Selling price	Tax payable
£12·90	£1·29	£14·19	£1·29

The total bill $= £14·19$

When goods pass through several hands, each seller adds VAT to the selling price, but he deducts from the sum that he is due to pay to the Customs and Excise any tax which he himself has paid. This avoids inflation of prices which would result from paying tax on a tax.

Example 2. A manufacturer sells 20 watches to a retailer for £100. The retailer sells the watches for £8 each, before the addition of VAT. How much tax does each pay, and what does a customer pay for one of the watches?

The retailer sells the watches for $20 \times £8 = £160$ (before VAT)

	Tax exclusive price (£)	VAT (£)	Selling price (£)	Tax to remit (£)
Manufacturer	100	10	110	10
Retailer	160	16	176	$16-10 = 6$
				$\overline{16}$

The customer pays $£8 + 80p = £8·80$ for a watch.

Notice that the total tax paid to Customs and Excise is 10% of the final tax-exclusive price of £160. This is true no matter how many hands the goods pass through.

The retailer's profit is £160 − £100 = £60

His profit per cent (based on his selling price) = $\frac{60}{160} \times 100 = 37\frac{1}{2}\%$.

Exercise 14

In this Exercise, assume that the rate of VAT is 10%, and round off answers to the nearest penny.

1 Calculate the total price paid by the customer (i.e. selling price + VAT) for each of the following.

   a a refrigerator at £52·50   b a television set at £287·20

   c a telephone bill for £8·32   d a lunch costing 85p

   e a man's suit paid for by a deposit of £6·50 and 24 weekly payments of 90p.

   f a hotel bill comprising:
      7 nights bed and breakfast at £2·50 per night
      4 lunches at 55p each, and 6 dinners at £1·04 each.

2 Calculate the tax remitted to the Government by the retailer in each of the following:

   a cost price = £485 + tax; selling price = £536 + tax
   b cost price = £1089 + tax; selling price = £1853 + tax
   c cost price = £48·50 + tax; profit of 20% on cost price.

3 Complete this table for a batch of shirts passing through four hands from the import of raw cotton to sale in a large shop.

Tax-exclusive price	VAT	Tax remitted
£4 500	...	...
£10 857	...	...
£18 486	...	...
£25 000	...	...

4 A manufacturer sells 40 clocks to a retailer for £800. The retailer sells them for £25 each (before VAT).

   a How much tax does each pay, and what does a customer pay for one of the clocks?

   b What is the retailer's profit per cent, based on the selling price?

5 Repeat question 4 for the case of 60 clocks for £350, retailing at £7·20 each.

Summary

1   Income and expenditure
(i) *Payslip*

GROSS PAY		DEDUCTIONS				NET
Basic wage	Overtime pay	Income tax	National insurance	Graduated pension	Super-annuation	PAY
38·00	11·40	2·82	0·88	1·85	2·00	41·85

(ii) *Foreign exchange*

$$\text{francs} \qquad\qquad £$$
$$11\cdot40 \longleftrightarrow 1$$
$$150 \longleftrightarrow 1 \times \frac{150}{11\cdot40}$$
$$= £13\cdot16$$

(iii) *Hire purchase*

Deposit	=	£6·50
12 instalments of £1·50	=	18·00
		24·50
Cash price	=	22·00
Extra cost by H.P.	=	£2·50

2   Savings and investment

e.g. Compound interest on £420 at 6% p.a.

P_1	=	£420·00
I_1	=	25·20
P_2	=	445·20
I_2	=	26·70
A	=	471·90
P_1	=	420·00
CI	=	£51·90

1% of P_1	=	£4·20
		6
6% of P_1	=	25·20
1% of P_2	=	4·45
		6
6% of P_2	=	26·70

3   *Insurance*

e.g. Premium on £7500 at £0·12%

Sum insured		Premium
£100	⟵⟶	£0·12
£7500	⟵⟶	$£0·12 \times \dfrac{7500}{100}$
		= £9

4   *Local rates*

e.g. Rateable value £120, rate 85p per £

Rateable value		Rates paid
£1	⟵⟶	£0·85
£120	⟵⟶	£0·85 × 120
		= £102

5   *Income tax*

Total income	=	£3500		Income tax at 30%
Allowances	=	£1526		= 30% of £974
Taxable income	=	£ 974		= £292·20

6   *Value added tax (VAT)*

	Tax exclusive price	VAT	Selling price	Tax to remit
Manufacturer	£100	£10	£110	£10
Retailer	£160	£16	£176	£6
				£16
				(= 10% of £160)

Statistics 3

1 Introduction

One of the aims in studying statistics is to analyse data, and this can be done in several ways. For example:

(i) *By representing the data in diagrams* by means of pictographs, pie charts, bar charts, line graphs, histograms and frequency polygons.

(ii) *By calculating measures of the 'average' and 'spread' of the data.*

Consider the two sets of marks:

$$13 \quad 14 \quad 15 \quad 16 \quad 17 \quad \text{and} \quad 5 \quad 10 \quad 15 \quad 20 \quad 25.$$

Each has an *average* of 15, but the *spread* of marks in the first is 17–13 i.e. 4, and in the second 25–5 i.e. 20. So measures of both *average* and *spread* are required in order to describe the data adequately.

This chapter falls therefore into three parts:

(i) Revision of diagrammatic representation of data.

(ii) Calculation of measures of central tendency—mean, median and mode.

(iii) Calculation of measures of spread or dispersion—range and semi-interquartile range.

2 Diagrams representing statistical data

(i) Pictographs and pie charts

Pictographs can provide a simple but striking display to illustrate certain data. They are more suitable for comparison than for measurement, and measurements should always be shown on the diagram, as in Figure 1.

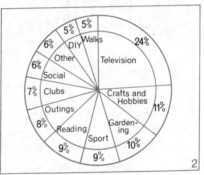

In a *pie chart* a circle is divided into sectors corresponding in size to the data involved; the area of each sector is proportional to the angle at the centre of the circle. Again, measurements should be shown, as in Figure 2.

(ii) Bar charts and line graphs

A *bar chart* like the one in Figure 3 has scales and titles clearly marked so that measurements concerning the data can be read from the chart.

Data which refer to information obtained over a period of time can be illustrated by means of a *line graph* where the horizontal axis shows the time scale, as in Figure 4. A pattern or trend can often be seen in such a graph, but care must be taken in deciding the meaning, if any, of intermediate points on the graph, or whether it is reasonable to extend the graph to the right to estimate future statistics. Even when possible, such 'interpolation' and 'extrapolation' provide estimates which must be treated with caution.

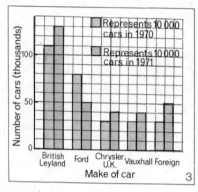

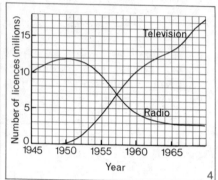

Exercise 1

1 The pictograph in Figure 1 depicts the sales of PRR Cat Food in 1961 and 1971. What is the actual increase in sales (by mass), and the percentage increase, between 1961 and 1971?

2 The pie chart in Figure 2 shows the percentage of time spent annually by a large sample of adults on leisure activities, according to a recent Government survey.

a What is the most popular leisure activity?

b What is the total percentage of time spent on crafts, sport and walking?

c If a person's leisure time in one year totals 1500 hours, how much time is spent on:

(1) reading (2) watching television?

3 The bar chart in Figure 3 shows the approximate number of cars sold by certain manufacturers in Britain in the first three months of 1970 and of 1971.

a Find the total number of these cars sold in each period of three months.

b Which firm had a serious strike in 1971 that affected production?

c Calculate the actual increase, and the percentage increase, between the two periods for sales of British Leyland, Vauxhall and foreign cars.

4   The line graphs in Figure 4 show the number of radio and television licences sold each year from 1945 to 1970.

a   In which year were most radio licences sold? How many?
b   In which year did the number of television licences equal the number of radio licences? What was the common number?
c   How many more television licences than radio licences were sold in 1965?
d   Describe, in a sentence or two, the general information provided by these graphs.

5   During an influenza epidemic the number of pupils absent in one school on ten consecutive school days was as follows:

Day	1	2	3	4	5	6	7	8	9	10
Number absent	30	35	60	78	87	83	69	43	30	29

a   Draw a *bar graph* of the distribution.
b   When was the epidemic at its worst?
c   Between which two days was there the greatest increase in absence?
d   For how many days were at least 50 pupils absent?
e   Calculate the mean number of pupils absent during these ten days.

6   The mass of a baby was recorded every fortnight for the first 16 weeks of its life, and the results are shown below:

Age in weeks	0	2	4	6	8	10	12	14	16
Mass in kg	3·2	3·3	3·6	3·9	4·1	4·1	4·4	4·9	5·3

a   Illustrate these data by a *line graph*.
b   During which fortnight was there   (1) the greatest   (2) the least increase in mass?
c   Have intermediate points on the line graph any meaning?
d   At what age would you estimate the child's mass to be 4·0 kg?

7   The number of vehicles per kilometre of road in certain countries in 1968 was estimated to be:

India	1	New Zealand	11	France	16	Netherlands	32
Australia	5	Japan	13	Switzerland	22	West Germany	32
Canada	10	USA	16	Italy	31	Great Britain	38

Illustrate these data in an appropriate diagram.

(iii) *Histograms and frequency polygons*

The frequency table below shows the number of fourth-year candidates, to the nearest 100, who obtained 0, 1, 2, 3, ..., 9 passes in Ordinary grade subjects in the 1971 Scottish Certificate of Education examinations:

Number of subjects passed	0	1	2	3	4	5	6	7	8	9
Number of candidates	1800	3300	4300	4700	4600	4700	5200	5500	3300	600

The table shows the *frequency* with which 0, 1, 2, 3, ..., 9 passes occurred, and so gives a *frequency distribution* of passes. Frequency distributions can be illustrated appropriately by means of *histograms* in which the areas of the rectangles are proportional to the frequencies. When the widths of the rectangles are equal, the heights of the rectangles are proportional to the frequencies. Figure 5 shows a histogram of the above frequency distribution of passes.

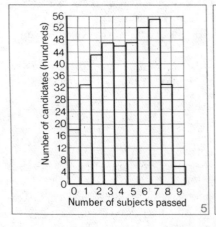 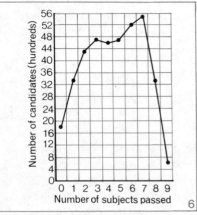

If the mid-points of the 'tops' of the rectangles in a histogram are joined we obtain a *frequency polygon*, as shown in Figure 6.

Exercise 2

1 Draw a histogram and frequency polygon to illustrate the frequency table below of the number of pupils, to the nearest 100, who ob-

tained 1, 2, 3, ..., 7 passes in Higher grade subjects in the 1971 S.C.E. examinations:

Number of subjects passed	1	2	3	4	5	6	7
Number of candidates	4900	4800	3900	3400	2900	1100	100

2 *a* Draw a histogram to illustrate this frequency table of the duration (to the nearest minute) of a sample of telephone calls through an exchange.

Duration (minutes)	0	1	2	3	4	5	6	7	8
Frequency	3	20	31	35	28	18	5	8	2

b How many calls lasted more than 5 minutes?
c What percentage of calls lasted less than 1 minute?
d Assuming that this sample is typical of calls through this exchange, what is the probability that a call chosen at random will last 3 minutes?

3 Pupils in the second year of a school measured the lengths of the middle finger of their left hands to the nearest 0·5 cm. The results were as follows:

Length in cm	8·0	8·5	9·0	9·5	10·0	10·5	11·0	11·5
Frequency	1	6	9	20	27	8	7	2

a Draw a histogram and frequency polygon on the same diagram.
b What percentage of the lengths are more than 10 cm?
c Assuming that this sample is typical of second-year pupils, what is the probability that a pupil chosen at random will have a finger of length 10 cm? At least 10 cm?

4 The rainfall (to the nearest cm) in 1971 at 110 stations in Scotland was as follows:

Rainfall in cm	50 –59	60 –69	70 –79	80 –89	90 –99	100 –109	110 –119	120 –129	130 –139	140 –149	150 –159
Frequency	2	18	28	15	17	12	4	8	3	2	1

a Illustrate these data as a frequency polygon.
b What is the relative frequency of rainfalls of 80–89 cm?
c What is the annual rainfall in your home area?

3 Measures of central tendency—mean, median and mode

For a collection of numbers or measures,

$$\text{the } mean = \frac{\text{the sum of all the measures}}{\text{the number of measures}}$$

the *median* is the middle measure when the measures are arranged in order

the *mode* is the most frequent measure.

Notation

If the symbol Σ (pronounced 'sigma') means 'sum of',

x represents one of the measures,

n is the number of measures,

then $\qquad$ mean $= \dfrac{\Sigma x}{n}$.

Example. Calculate the mean, median and mode of:

$$8, 9, 9, 9, 10, 11, 12, 12$$

$$\text{Mean} = \frac{\Sigma x}{n} = \frac{80}{8} = 10$$

$$\text{Median} = \frac{9 + 10}{2} = 9\cdot5$$

$$\text{Mode} = 9$$

Exercise 3

Find the mean, median and mode of each of the following (to 1 decimal place where necessary):

1   2, 3, 4, 4, 7

2   2, 3, 4, 5, 5, 6, 7, 8

3   2, 3, 4, 4, 4, 5, 6, 6, 6, 6, 7

4   8, 9, 7, 8, 5, 6, 9, 7, 9, 10, 9

5   1, 5, 4, 2, 1, 1, 3, 5, 4, 3, 2, 4, 6, 4

6   £12, £13, £13, £13, £14, £15, £15, £16, £17,

4 Calculating the mean of a frequency distribution

We can calculate the mean of the frequency distribution of passes given on page 200 from the following table:

Number of subjects passed (x)	Number of candidates, i.e. frequency (f)	Total number of passes (fx)
0	1800	0
1	3300	3300
2	4300	8600
3	4700	14 100
4	4600	18 400
5	4700	23 500
6	5200	31 200
7	5500	38 500
8	3300	26 400
9	600	5400
	$\Sigma f = 38\,000$	$\Sigma fx = 169\,400$

$$\text{Mean} = \frac{\Sigma fx}{\Sigma f} = \frac{169\,400}{38\,000} = 4 \cdot 5$$

Class intervals

We saw in the Statistics chapter in Book 4 that in the case of a large number of measures it is usual to group them in *class intervals*. The table below lists the diameters, to the nearest millimetre, of a batch of ball-bearings produced by a machine:

78	72	74	74	79	71	75	74	72	68
72	73	73	72	74	75	74	74	73	72
66	75	74	73	74	72	79	71	70	75
80	69	71	70	70	80	75	76	77	67

We can group these in equal *class intervals*; intervals of 2 mm or 3 mm would be appropriate. We can now calculate the mean diameter more easily, but at the expense of precision. Sometimes the only information available is the frequency for each class interval.

We then have to assume that the frequencies are spread evenly over each interval, which comes to the same thing as assuming that they are all concentrated at the *midpoint of the interval*. Choosing a class interval of 3 mm, we have:

Diameter in mm	Midpoint (x)	Frequency (f)		fx
65–67	66	||	2	132
68–70	69	₩	5	345
71–73	72	₩ ₩ |||	13	936
74–76	75	₩ ₩ ||||	14	1050
77–79	78	||||	4	312
80–82	81	||	2	162
			$\Sigma f = 40$	$\Sigma fx = 2937$

Mean diameter $= \dfrac{\sum fx}{\sum f} = \dfrac{2937}{40} = 73$ mm, to two significant figures.

Note on class limits and class boundaries

In the above distribution the modal class interval is 74 mm–76 mm; 74 mm and 76 mm are called the *class limits* of the interval.

Before rounding off, some diameters may lie outside the class limits, e.g. 76·2 mm. Where this matters, we define *class boundaries* 73·5 mm and 76·5 mm for the modal class above.

For the class interval 74 mm–76 mm we have:

a *lower class limit* of 74 mm and an *upper class limit* of 76 mm;

a *lower class boundary* of 73·5 mm and an *upper class boundary* of 76·5 mm.

Then every measure x would be contained within class boundaries as follows:

$$64\cdot5 \leqslant x < 67\cdot5; \quad 67\cdot5 \leqslant x < 70\cdot5; \quad ...; \quad 79\cdot5 \leqslant x < 82\cdot5.$$

Exercise 4

1 Paper clips are sold in boxes which normally contain 100 clips. The numbers of clips in a random sample of 150 boxes are shown in Table 1.

a Construct a frequency table with the headings 'number of clips (x)', 'frequency (f)' and 'total number of clips (fx)'. Hence calculate the mean number of clips in a box.

b If this is a typical sample of boxes, what is the probability that if a box is chosen at random it will contain (1) 97 clips (2) more than 99 clips?

2   The lengths, to the nearest centimetre, of fish caught in a competition are shown in Table 2. Construct a frequency table, and calculate the mean length of fish caught.

	TABLE 1		TABLE 2		TABLE 3
Number of clips	Frequency	Length in cm	Frequency	Reading score	Frequency
95	1	48	1	0–4	15
96	5	49	1	5–9	60
97	10	50	1	10–14	125
98	22	51	0	15–19	260
99	30	52	2	20–24	250
100	37	53	2	25–29	200
101	20	54	1	30–34	90
102	15	55	1		
103	10	56	0		
104	0	57	5		
		58	6		
		59	10		
		60	5		
		61	3		
		62	2		

3   In a survey of reading ability the scores obtained are shown in Table 3.
a   Illustrate these data in a histogram.
b   What is the modal class?
c   Calculate the mean reading score.

4   Calculate the mean of the following numbers:

a   without grouping them
b   grouping them in class intervals 0–2, 3–5, ..., 18–20.

8	12	15	10	11	11	16	9	5	8	11	3
2	12	20	10	9	15	5	17	17	6	14	8
18	8	13	16	19	5	7	20	19	8	9	12

5   The lengths, to the nearest millimetre, of a sample of a certain type of leaf are given below. Calculate the mean length:

a   without grouping the data
b   by grouping the data in class intervals 47–49, 50–52, etc.

| 53 | 51 | 58 | 54 | 59 | 60 | 52 | 52 | 55 | 49 | 51 | 53 |
| 55 | 60 | 58 | 57 | 51 | 57 | 56 | 50 | 53 | 58 | 59 | 57 |

6   Calculate the mean of the following scores:

a   without grouping the data
b   grouping the data in class intervals 40–49, 50–59, etc.

44	54	85	62	73	57	99	91	66	74
83	49	57	52	64	67	73	82	90	70
48	58	65	76	88	90	75	68	77	62

7   The weekly earnings (in £) of 50 men in a factory are given below. Choose an appropriate class interval and calculate the mean earnings.

27·50	28·00	26·95	30·85	31·00	28·50	28·80	29·25	27·85	28·75
26·35	26·35	26·75	28·00	28·00	28·65	29·25	27·75	27·75	28·25
27·65	27·85	26·90	27·00	27·00	27·50	28·25	29·50	28·25	28·00
30·00	29·20	29·20	30·00	29·00	28·20	28·80	27·65	29·65	28·00
29·50	30·00	26·80	27·25	27·50	27·50	28·00	28·50	27·50	30·25

---

5 Alternative methods for calculating the mean

(i) Using an assumed mean

We start by estimating the mean, and then calculate the correction to be applied to this by finding the mean of the deviations from the assumed mean, as shown below.

Example 1. Find the mean of the nine ages listed (it being understood that 14 years 8 months means up to, but not including 14 years 9 months, as is common practice in recording ages).

Age		Assumed mean		Deviation
years	months	years	months	(months)
14	8			−4
14	9			−3
14	9			−3
14	10			−2
15	0	15	0	−12
15	0			0
15	1			+1
15	1			+1
15	2			+2
				+4

Sum of deviations $= +4-12 = -8$ months.

The mean deviation $= -\frac{8}{9}$ month.

Mean age $= 15$ years 0 months $+(-\frac{8}{9})$ month

$= 14$ years 11 months.

Note. Before summing the deviations, work may be saved by cancelling corresponding positive and negative deviations (in the above example, $-4+1+1+2 = 0$).

Example 2. Find the mean of the following frequency distribution of marks:

Score	Midpoint of interval	Frequency (f)	Deviation (d) (assumed mean $= 67$)	fd
50–54	52	4	-15	-60
55–59	57	6	-10	-60
60–64	62	8	-5	-40
				-160
65–69	67	16	0	
70–74	72	10	5	$+50$
75–79	77	3	10	$+30$
80–84	82	2	15	$+30$
85–89	87	1	20	$+20$
		$\underline{50}$		$\underline{+130}$

Sum of deviations $= 130 - 160 = -30$

The mean deviation $= -\dfrac{30}{50} = -0\cdot6$

Mean score $= 67 - 0\cdot6 = 66\cdot4$

(ii) Using a calculating machine

The calculation of the mean diameter of the ball-bearings in the example of Section 4 can be carried out quickly by machine. If you set 1 at the left-hand end of the setting register and the midpoint values at the right-hand end, the products can be summed as you proceed. You will finish with the sum of the frequencies at the left and the sum of the products at the right of the product register. The division can then be done in the usual way.

Exercise 5B

1–6 Using either of the above methods, calculate the mean for each of the distributions given in Exercise 7, questions 2 to 7 on page 211.

---

6 Measures of spread or dispersion

(*i*) Quartiles

The median divides a collection of ordered measures into two equal parts. The quartiles divide the measures into four equal parts. If the measures are arranged in order on a line, as shown in Figure 7, Q_1 is the *lower quartile*, Q_2 is the *median* and Q_3 is the *upper quartile*.

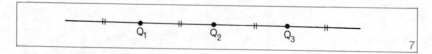

7

Example 1.

2	(3)	4	(6)	8	(9)	11
	Q_1		Q_2		Q_3	

Fix the position of Q_2 first, then Q_1 and Q_3.
Here $Q_1 = 3$, $Q_2 = 6$ and $Q_3 = 9$.

Example 2.

2	(3)	4	↑	6	(8)	9
	Q_1		Q_2		Q_3	

Here $Q_1 = 3$, $Q_2 = 5$ and $Q_3 = 8$.

Example 3.

2	3	↑	4	6	↑	8	9	↑	11	14
		Q_1			Q_2			Q_3		

Here $Q_1 = 3 \cdot 5$, $Q_2 = 7$ and $Q_3 = 10$.

Exercise 6

1 Find the lower quartile (Q_1), the median (Q_2) and the upper quartile (Q_3) for each of the following sets of data:

a 5, 6, 7, 7, 8, 10
b 28, 28, 29, 31, 32, 32, 34, 35, 37, 37, 37, 39
c 101, 104, 106, 107, 109, 109, 111, 113, 113, 115, 116, 118, 118, 118, 120, 121, 125
d 15, 13, 7, 16, 11, 10, 13, 9, 16, 8, 10

2 The duration, in minutes, of telephone calls made by a business man on a certain day was:

| 3 | 7 | 10 | 2 | 4 | 8 | 11 | 9 | 6 | 3 |
| 2 | 4 | 8 | 15 | 14 | 10 | 8 | 7 | 4 | 7 |

Calculate the median and quartiles.

3 The number of hours of sunshine recorded daily in a city during a fortnight in April 1973 was as follows:

| 3·8 | 7·8 | 5·7 | 2·0 | 3·4 | 7·2 | 4·1 |
| 4·8 | 6·3 | 0·8 | 1·3 | 7·9 | 7·6 | 5·2 |

Calculate the median and quartiles.

4 Labels on the tins of a certain make of orange juice state that the volume of the contents is 0·50 litre. Find the median and quartiles for the volumes (given to the nearest 0·01 litre) in the following sample:

| 0·52 | 0·51 | 0·53 | 0·52 | 0·52 | 0·50 | 0·50 | 0·54 | 0·51 |
| 0·51 | 0·49 | 0·53 | 0·54 | 0·51 | 0·49 | 0·50 | 0·53 | 0·51 |

(ii) Cumulative-frequency table and curve

Sometimes we can analyse the data in a frequency distribution more easily if we construct a *cumulative-frequency distribution table* and *curve* as follows. The table contains the marks obtained by 200 pupils in an examination.

Marks	Frequency	Cumulative frequency
1–10	7	7
11–20	9	16
21–30	12	28
31–40	21	49
41–50	39	88
51–60	44	132
61–70	29	161
71–80	18	179
81–90	14	193
91–100	7	200
	200	

The cumulative-frequency column shows that:

7 pupils scored 10 marks or less
16 pupils scored 20 marks or less
28 pupils scored 30 marks or less
....
there were 200 pupils altogether.

These facts can be illustrated by graphing the cumulative frequency against the upper class limits, as shown in Figure 8. In theory, upper class boundaries (as defined on page 204) should be used in the case of continuous variables such as length, mass, time, etc.; but in practice it is simpler and sufficiently accurate to use upper class limits in these *cumulative-frequency graphs.*

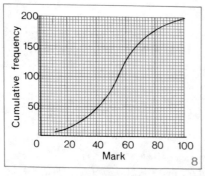
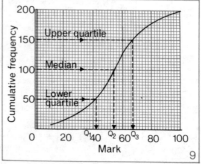

The cumulative-frequency curve is sometimes known as an *ogive* from a curve used in architecture.

The quartiles are found from the cumulative-frequency curve as shown in Figure 9. Here $Q_1 = 41$, $Q_2 = 53$ and $Q_3 = 66$. This means that 50 pupils scored up to 41 marks, 100 scored up to 53 marks and 150 scored up to 66 marks.

Exercise 7

1 Figure 10 shows two cumulative-frequency curves. Estimate the median, the lower quartile and the upper quartile of each distribution.

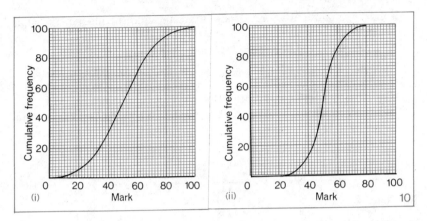

(i)   Mark

(ii)   Mark   10

In the following questions, plot the cumulative frequency against the upper class limit, using 2-mm squared paper.

2   Table 1 gives the frequency of marks 21–30, 31–40, etc., in an examination. Construct a cumulative-frequency table and curve. Hence estimate:

a   the median and quartiles

b   the number of pupils who pass if the passmark is 53.

c   the passmark if two-thirds of the pupils pass.

TABLE 1		TABLE 2		TABLE 3	
Mark	Frequency	Weekly takings (£)	Frequency	Age of machine (months)	Number of machines
21–30	1	480–499	2	0–5	3
31–40	1	500–519	3	6–11	7
41–50	3	520–539	5	12–17	10
51–60	9	540–559	7	18–23	19
61–70	8	560–579	11	24–29	27
71–80	6	580–599	13	30–35	31
81–90	2	600–619	6	36–41	16
		620–639	4	42–47	9
		640–659	0	48–53	6
		660–679	1	54–59	2

3   The weekly takings of a shopkeeper in one financial year are classified in Table 2. Draw up the cumulative-frequency table and curve, and estimate the median and quartiles.

4   Table 3 gives the ages in months of machines in a factory.

    a Draw a cumulative-frequency curve, and find the median and quartile ages.

    b How many machines are less than 20 months old?

5   Table 4 lists the ages of first-year pupils in a school.

    a Find their median and quartile ages.

    b If it is required to identify the youngest 100 pupils, what ages would be included?

TABLE 4

Age in years and months	Frequency
12,0–12,2	10
12,3–12,5	24
12,6–12,8	36
12,9–12,11	53
13,0–13,2	46
13,3–13,5	32
13,6–13,8	15

TABLE 5

Diameter in mm	Frequency
140–142	3
143–145	7
146–148	21
149–151	48
152–154	19
155–157	2
158–160	0

TABLE 6

Mark	Frequency 1st term	2nd term
31–40	1	1
41–50	7	1
51–60	8	3
61–70	5	11
71–80	3	8
81–90	1	1

6   A machine was set to produce metal rods of diameter 150 mm. To test its accuracy a sample of 100 rods was taken, and their diameters were measured to the nearest mm, giving Table 5. Construct a cumulative-frequency table and curve. Estimate the limits between which the lengths of the middle 50 rods must lie.

7   Table 6 shows the frequency distribution of marks obtained by a class in two examinations.

    a Draw the cumulative-frequency curves for both examinations.

    b Estimate the median and quartile marks for each examination.

    c In each case find the passmark if 60% of the pupils have to pass.

(iii) Range and semi-interquartile range

We have seen how the different 'averages'—mean, median and mode—all serve useful purposes when we are investigating statistical data, by giving us measures of central tendency.

    But we have still to devise a measure of the *spread*, or *dispersion*, of the data. Each of the following sets of marks has a mean of 57, but obviously the spread of marks is different in each case.

English	45	48	53	53	62	68	70
Mathematics	18	37	50	60	69	73	92
French	10	47	58	64	67	75	78

To compare the spread, or dispersion, of marks we can use:

(i) The *range* = highest measure − lowest measure.

For English the range is $70 - 45 = 25$, for Mathematics 74, for French 68.

Although easy to calculate, the range is affected by the presence of a freak mark at either end which distorts the general picture; for example, the mark of 10 in French.

(ii) The *semi-interquartile range* $= \frac{1}{2}(Q_3 - Q_1)$.

For English the semi-interquartile range $= \frac{1}{2}(68 - 48) = 10$, for Mathematics $\frac{1}{2}(73 - 37) = 18$, for French $\frac{1}{2}(75 - 47) = 14$.

To summarise, we have

Subject	Mean	Range	Semi-interquartile range
English	57	25	10
Mathematics	57	74	18
French	57	68	14

Consider now the more realistic case where a larger number of data is involved. The table below gives the frequency of marks scored in Science and Latin by a group of 150 pupils.

Mark	1–10	11–20	21–30	31–40	41–50	51–60	61–70	71–80	81–90	91–100
Frequency Science	2	4	7	16	34	39	24	13	9	2
Latin	0	0	1	16	46	48	27	11	1	0

The means are almost identical, 53·1 and 53·2 respectively, but it is obvious from the table that the marks in Latin are clustered closely together, while those in Science are much more spread out or dispersed. This difference in spread or *dispersion* is even more apparent if we draw the frequency polygons, as in Figure 11.

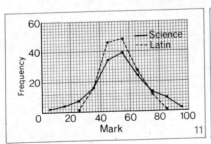

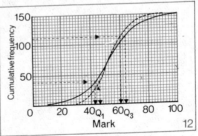

The cumulative-frequency curves for the distributions are shown in Figure 12. You will see that the middle portion of the curve is appreciably steeper in the case of Latin, indicating that there is a larger number of pupils scoring between 40 and 70 in Latin than in Science.

From the graphs we obtain the following results:

		Science	Latin
Upper quartile	Q_3	64	60
Lower quartile	Q_1	43	46
Semi-interquartile range	$\frac{1}{2}(Q_3 - Q_1)$	10·5	7

Thus the dispersion of the marks in Science is one and a half times that of the Latin marks.

Consider now the case of a boy who scores 80% in both Latin and Science. These are good marks, but are they equally good in both subjects? To answer this we look at the graphs and find that 80% in Latin corresponds to a cumulative frequency of 149, while 80% in Science corresponds to a cumulative frequency of 139. Hence the boy is 2nd in Latin but 12th in Science.

Experiment.—Take a sheet of 2-cm squared paper about 50 cm square, and lay it flat. Select a 'target square' near the centre and drop about 100 grains of rice, 2 or 3 at a time, on to the target from a height of about 30 cm.

Write in each square the number of grains it contains. Choose an origin and axes, and construct frequency tables of the number of grains in the 2-cm columns parallel to: *a* the *y*-axis *b* the *x*-axis.

Calculate the mean distance, from the origin or from the target area, of the grains.

Illustrate the distribution by means of a histogram and a cumulative-frequency diagram. Estimate the median and quartiles of the distribution.

By dropping the rice from a greater height, notice the differences in the means and semi-interquartile ranges in the two distributions.

Exercise 8

1   Calculate the range and semi-interquartile range for the numbers:

53, 54, 57, 59, 62, 63, 65, 69, 69, 71, 72, 75, 78, 78, 80, 83.

2–5 Calculate the range and semi-interquartile range for the information given in questions *1–4* in Exercise 6 on page 208.

6   Table 1 gives a distribution of expenditure on school supplies in 1970–71 for the 36 cities and counties of Scotland. Draw a cumulative-frequency curve, and find the median expenditure per pupil and the semi-interquartile range.

| TABLE 1 | | TABLE 2 | |
Expenditure (£)	Frequency	Number of hours sunshine	Number of stations
4–4·90	5	800–899	1
5–5·90	7	900–999	2
6–6·90	12	1000–1099	2
7–7·90	8	1100–1199	3
8–8·90	1	1200–1299	25
9–9·90	2	1300–1399	27
10–10·90	1	1400–1499	5
		1500–1599	4
		1600–1699	3

7   The number of hours of sunshine recorded at 72 meteorological stations in Britain in 1971 is given in Table 2. Find the median number of hours of sunshine, and the semi-interquartile range of the distribution.

8   Table 3 gives the English and Mathematics marks for a group of pupils.

a Draw cumulative-frequency curves, and find the median mark and semi-interquartile range for each distribution.

b Calculate the mean mark for each distribution.

c Describe the distributions in a sentence or two.

| TABLE 3 | | | TABLE 4 | |
| | Frequency | | Number of letters per word | Frequency |
Mark	English	Mathematics		
21–30	0	2	1	50
31–40	1	3	2	120
41–50	4	6	3	184
51–60	20	10	4	255
61–70	14	12	5	173
71–80	8	10	6	105
81–90	2	4	7	63
91–100	1	3	8	40
			9	10

9 In a study of word-lengths used by an author, 1000 words were chosen at random from his writing, and the distribution in Table 4 was obtained. Analyse the data by finding:

a the mode, mean and median of the distribution

b the range and semi-interquartile range.

Summary

1 *Diagrams representing statistical data include:*

pictographs, pie charts, bar charts, line graphs, histograms and frequency polygons.

2 *Measures of central tendency:*

(i) $Mean = \dfrac{\text{the sum of all the measures}}{\text{the number of measures}} = \dfrac{\sum x}{n}$

(ii) *Median* is the middle measure in a collection of ordered measures.

(iii) *Mode* is the most frequent measure.

3 *Calculating the mean of a frequency distribution*

(i)

Measures	Midpoint (x)	Frequency (f)	fx
2–4	3	2	6
5–7	6	22	132
8–10	9	10	90
		$\sum f = 34$	$\sum fx = 228$

$$Mean = \frac{\sum fx}{\sum f} = \frac{228}{34} = 6.7$$

[OVER

(ii)

Measures	Midpoint (x)	Frequency (f)	Deviation (d) Assumed mean $= 6$	fd
2–4	3	2	-3	-6
5–7	6	22	0	0
8–10	9	10	3	30
		$\Sigma f = 34$		$\Sigma fd = 24$

$$\text{Mean} = 6 + \tfrac{24}{34} = 6{\cdot}7$$

4   Measures of spread, or dispersion

The *quartiles* Q_1, Q_2 (the median) and Q_3 divide the ordered measures into four equal parts. They are obtained from a *cumulative-frequency curve*.

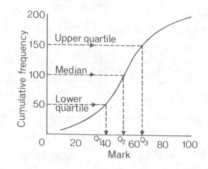

The *range* = highest measure − lowest measure.

The *semi-interquartile range* $= \tfrac{1}{2}(Q_3 - Q_1)$.

Topic to explore

Permutations

(i) Permutations, by arrays and tree diagrams

When studying probability in Book 3 we had to find out the number of *possible outcomes* of an experiment. For example, in throwing a die there are 6 possible outcomes, namely 1, 2, 3, 4, 5 and 6.

Example. A first and a second prize are to be given for a race in which Alex, Bill, Charles and Dave take part. In how many ways can the prizes be awarded?

The *possible outcomes* of the race can be set out as follows:
(i)   as an *array* of ordered pairs:

AB	BA	CA	DA
AC	BC	CB	DB
AD	BD	CD	DC

There are 12 possible outcomes, so the pair of prizes can be awarded in 12 different ways.

(ii)   as a *tree diagram*:
The first-prize placings are shown at the ends of the first branches from the left, and the second-prize placings at the ends of the right-hand branches.

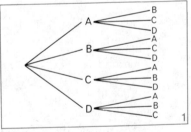

The ordered pairs can be read from the diagram.

In this example we have found all the possible *arrangements* or *permutations* of four boys, taken two at a time.

Exercise 1

Set out the elements in the following examples as arrays, and/or tree diagrams; it may be helpful to name the members of each team A, B, C and so on.

1   Find the number of ways in which a captain may be chosen from:

  a a netball team of 8 players   *b* a hockey team of 11 girls
  c a rugby team of 15 boys.

2   Find the number of ways in which a captain and a vice-captain may be selected in: *a* a chess team of 4 players   *b* a tennis team of 6 players.

3   In how many ways can a chairman and secretary be chosen in a committee of 2 members? Repeat the question for committees of 3 and 4 members.

4   Four items have to be placed in order in a competition. In how many ways can we place: *a* the first   *b* the first and second   *c* the first, second, and third   *d* all four items?

5   How many different two-digit numbers can be formed from elements of the set $\{1, 2, 3, 4, 5, 6\}$, assuming that no repetition of digits is allowed in any of the numbers?

(ii) Permutations, by 'thinking it out'

Can you answer question **5** above without using an array or tree diagram? Do you see that the first digit can be chosen in 6 ways? For each of these 6 ways, how many ways are there of choosing the second digit? How many ways are there of choosing the first and *then* the second digit?

Do you see that there must be 6×5, or 30, permutations for the first two digits?

In how many ways could a third digit now be chosen? So how many three-digit numbers are there?

Suppose we wanted to find the total number of permutations of all 6 digits.

We can think in terms of filling the spaces in Figure 2.

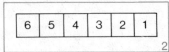

6	5	4	3	2	1

2

The first space can be filled in 6 ways, the second in 5, the third in 4 ways, and so on. The total number of permutations is:

$$6 \times 5 \times 4 \times 3 \times 2 \times 1 = 720.$$

Note. $6 \times 5 \times 4 \times 3 \times 2 \times 1$ is sometimes written $6!$ or $\underline{|6}$, and read as '6 factorial' or 'factorial 6'.

Exercise 2

1   How many ways are there of choosing a school captain and a vice-captain from 25 prefects?

2   How many permutations can be obtained by choosing 2 elements, in order, from a set of: *a* 10 elements   *b* 20 elements   *c* 100 elements   *d* 1001 elements?

3   How many different arrangements can be obtained by drawing 2 cards, in order, from a pack of 52 playing cards?

4   In a magazine competition the following items have to be arranged in order of 'desirability': radio, TV set, washing machine, refrigerator, dishwasher, central heating, fitted sink unit, fitted carpets. How many permutations are possible?

5   There are 5 different routes from Lenzie to Bishopbriggs, and 4 different routes from Bishopbriggs to Glasgow city centre. How many different routes are there from Lenzie to Glasgow city centre via Bishopbriggs?

6   Six photographs were entered for a competition. In how many different ways can we choose the first, second, and third?

7   At an exhibition we have to place paintings in order in a row from left to right. In how many different ways can they be arranged if there are: *a* 3 paintings   *b* 4 paintings   *c* 5 paintings   *d* 6 paintings?

8   A 'combination' lock for a bicycle has three rings each marked with the numerals 0, 1, 2, ..., 9. How many different arrangements of the numerals are possible on the lock? Assuming that only one of these unlocks the bicycle, how many wrong guesses might a bicycle thief have to try before managing to steal the bicycle?

9   How many different permutations can we obtain when 2 dice are thrown, one after the other?
      What is the probability of throwing a double?

10a   In tossing a penny, how many outcomes are there?

  b   In tossing 2 pennies, one after the other, how many outcomes are there?

  c   What is the total number of permutations obtained by tossing 3 pennies, one after the other?

  d   Continue for the cases of 4, 5 and 6 pennies. Can you write down the result for n pennies?

  e   Show the results for 4 pennies as a tree diagram.

  f   If n pennies are tossed, one after another, what is the probability that: (1) n heads (2) either n heads or n tails (3) $(n-1)$ heads and 1 tail, turn up?

11   A football match can result in a home win, an away win, or a draw. How many ways are there of forecasting the results of:
      a   one match   b   two matches   c   three matches?
          It is interesting to multiply out the number of ways of forecasting a list of fourteen matches. Try it, and realise that only one of these forecasts can be correct. What is the probability of forecasting the 'winning line'? Many problems on permutations appear to be connected with gambling; but rather than encouraging you to gamble, your study of this subject should make you appreciate how small the chances of winning are.

Revision Exercises

Revision Exercise on Chapter 1
Social Arithmetic 2

Revision Exercise 1

1 A firm decides to send out 10 000 advertising circulars. The cost of printing is £6·75 for the first 5000 and £1·05 for each additional 1000. Postage costs $2\frac{1}{2}$p for each item. Find the total cost of the advertising campaign.

2 A factory which uses 30 tonnes of coal weekly, costing an average of £18·50 per tonne, is considering changing over to electricity, and is quoted a cost of 1p per unit. How many units could be obtained per week for the cost at present of the coal?

3 A camera is priced 320 francs in France. How much will it cost me in £s to buy it and take it to Britain, assuming that the Customs charge 35% of the cost as duty, and that £1 $\doteqdot$ 12 francs?

4 A travel agency quotes £59·40 for a holiday in Majorca for a person travelling from London by air. There is a supplement of £12·25 per person for flying from Aberdeen. Find the total cost of the holiday for Mr and Mrs Brown of Aberdeen who spent 10 400 pesetas while in Majorca. (£1 = 150 pesetas.)

5 A portable typewriter is advertised for £15·20 cash or 12 monthly payments of £1·44. How much extra is paid in the latter case, and what percentage is this of the cash price?

6 Find the total amount in my account in the Savings Bank at the end of 11 months if I deposited £500 at the beginning of the period, and the rate of interest is 8% per annum.

7   If depreciation is reckoned at the rate of 20% per annum, calculated on the value at the end of the previous year, how long is it worthwhile keeping a piece of machinery costing £1200 if its scrap value is taken to be £450? (Work to the nearest £.)

8   A man borrows £4000 from a Building Society. He pays back £564 at the end of the first year, which covers an interest charge of $10\frac{1}{2}\%$ on the loan as well as reducing his debt. How much does he owe the Society at the end of the year?

9   I borrow £1000 and promise to repay £400 at the end of the first year and £500 at the end of the second year. Interest is added at 10% per annum at the end of each year on each complete £ outstanding in that year. What must I pay at the end of the third year to clear the debt?

10   A man takes out a 'with profits' assurance policy for £3000. The monthly premium is £0·45 per £100 assured. Find the total amount paid by the man over a period of 20 years.

   If, when the policy matures, the overall 'profit' paid out on the policy is 35% of the sum assured, find the overall 'profit' or 'loss' on money invested in this policy.

11   A man aged 40 years takes out an assurance policy for £2000 payable in 20 years. If the premium is £7·50 per month, find the total cost of the policy. If he is allowed income tax relief at the rate of 30% on one half of his premium, how much does the policy actually cost him annually?

12   Calculate the missing quantities in the following table:

	Rateable value	Rate per £	Amount paid in rates
a	£95	£0·80	...
b	£120	...	£72
c	£320	£0·74	...
d	£5·6 million	...	£4·7 million

13   Two brothers, James and John, own houses in different towns. James's house has a rateable value of £126 and is charged rates at £0·80 in the £. John's house has a rateable value of £120 and he pays exactly the same total sum of money in rates as James does. Find the rate per £ charged in the town where John lives.

14 Using the information about allowances and income tax given on page 189 calculate how much tax should be deducted monthly from the salaries of the following men:

a a single man earning £1800 a year
b a married man, with 3 children aged 5, 9 and 14 years, who earns £2345 per year
c a single man with a salary of £4800 per annum, and life assurance with a premium of £600.

15 Calculate the compound interest on:

a £2000 for 3 years at 4% per annum
b £250 for 2 years at 5% per annum.

16 Calculate the total price paid for dinner by a party of four if the meal is priced at £1·95 each and coffee costs 15p per person. (Total price paid = list price + VAT at 10%).

Revision Exercise on Chapter 2
Statistics 3

Revision Exercise 2

1 The percentages of people in various occupations in a town are as follows: professional, 3%; managerial, 14%; skilled, 51%; semi-skilled, 15%; unskilled, 17%. Illustrate this information as a bar chart and as a pie chart.

2 The number of megawatts of electrical power supplied by Portobello power station in Edinburgh on Christmas Day 1972 every half-hour from 00 30 onwards was as follows (reading along each row in turn):

76	70	61	49	44	40	38	37	36	36	36	36	36	39	40	43
52	63	78	88	94	97	98	97	96	93	92	93	89	86	86	86
82	76	71	72	70	70	72	75	75	75	74	74	78	84	81	74

Illustrate the above by means of a line graph. Describe the graph in a sentence or two, suggesting reasons for its shape.

3   The table shows the times in seconds (to the nearest tenth) for rounds in a competition at the 'Horse of the Year' show:

61·4	38·8	36·8	40·2	53·9	45·0	42·1	37·7	50·7	36·4
42·4	48·0	45·2	49·2	47·2	49·7	49·8	35·6	37·7	37·6
36·7	48·9	51·8	41·5	41·0	40·2	49·6	59·6	36·7	37·5

Construct frequency tables, using class intervals of:
a 2 seconds   *b* 3 seconds   *c* 5 seconds, each starting at 35 seconds. Draw the corresponding histograms.
Which class interval do you consider to be most suitable?

4   Use one of the frequency tables in question *3* to calculate the mean time in the competition rounds.

5   In a recent experiment a group of pupils sat two examinations of different types. The results were as follows:

Type A

44	50	43	45	43	43	46	35	77	51	43	69	70	71
70	62	51	55	44	54	53	47	42	52	56	35	50	

Type B

83	88	89	66	65	76	77	58	48	41	47	74	58	61
52	54	46	46	56	44	50	42	46	43	45	50	36	

Group each of these sets of data (class intervals 35–39, etc.) and illustrate by histograms. Calculate the mean of each distribution.

6   Using the frequency tables you made in question *5*, draw cumulative-frequency curves. Estimate the median and semi-interquartile range for each distribution.

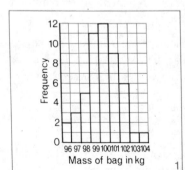

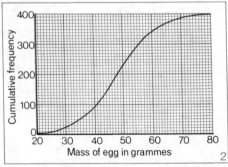

7   A machine is used to fill bags with flour and is set so that each bag should contain approximately 100 kg. A sample run of bags is taken and each bag weighed to the nearest kg. The results are shown in Figure 1. Use the information to calculate the mean mass of the sample, to draw a cumulative-frequency curve and to find the semi-interquartile range of the masses.

8   400 eggs sent as a sample to a packing station were weighed in grammes, and the results are shown in the cumulative-frequency curve in Figure 2. Eggs weighing less than 30 g are sent for processing. The rest are graded thus: small, 30 g but less than 45 g; medium, 45 g but less than 60 g; large, 60 g and over.

a   What percentages are sent for processing, graded small, graded medium and graded large?
b   What is the probability that an egg selected at random from this typical sample weighs   (1)   less than 55 g   (2)   more than 65 g?
c   Estimate the median and semi-interquartile range of the masses.

9   In a new town there are three types of house, described as 'Houses', 'Flats', and 'Managerial houses'. A census was taken of the number of persons per house, and the results are given below:

Number of persons per house		1	2	3	4	5	6	7	8	9
Frequency:	Houses	6	236	553	859	350	120	44	18	9
	Flats	87	415	335	51	6	0	0	0	0
	Managerial	1	8	10	17	9	2	0	0	0

a   Draw a histogram and frequency polygon for each distribution.
b   Calculate the mean number of persons in each type of house.

10   Out of every 100 families coming to live in the new town in each of the years 1960 and 1970, the number of families with given numbers of children is shown below:

Number of children	0	1	2	3	4	5	6	7
Number of families:								
arriving in 1960	5	23	32	23	11	3	2	1
arriving in 1970	25	28	30	12	3	1	0	1

a   Draw a histogram and cumulative-frequency curve for each distribution.
b   Calculate means, modes, medians, and semi-interquartile ranges for the two distributions.

c   Can you draw any conclusion, from the figures, about a change in the kind of families coming to the town in these two years?

11   Select a passage containing about 300 words from one of your English books, and record the frequency of the use of the various letters of the alphabet. Express the frequencies of, say, the six most frequently used letters as percentages of the total number of letters in your sample. Compare your results with those of other members of your class who have chosen different passages. Can you see how results of this kind can be used for 'breaking' codes?

Repeat this investigation using passages from other languages, e.g. French, German, Latin, and consider whether it appears that the same letters are the most frequently used ones.

Cumulative Revision Section (Books 1-6)

Cumulative Revision Exercise

1   Which is the greater, (1) or (2) in each case?

   a (1) $0.65 \times 0.08 \times 20$,   or   (2) $1.6 \div 1.5$

   b (1) $\frac{3}{4}$ of £1·56,   or   (2) $\frac{5}{8}$ of £1·84

   c (1) 125% of 64,   or   (2) 64% of 125

2   Calculate the total cost of:

   1000 copies of *Modern Mathematics* at £1·20 each
    700 copies of *Three-figure Tables* at 16p each
   1800 Exercise books at £3·25 per 100.

3   Calculate the following, rounding off *a* and *b* to 1 decimal place, and *c* and *d* to 2 significant figures:

   a $29.38 + 36.79 - 41.83$ *b* $1.63 \div 2.3$ *c* $\sqrt{234}$ *d* $\sqrt{0.91}$

4   Calculate the electricity bill for meter readings of 7059 and 8611 if the charges are 3p per unit for the first 36 units, 0·8p per unit for the next 100 units and 0·5p per unit for the remainder.

5   Write down 3 terms which could follow in each of these sequences:

   a 7, 11, 15, 19, ...   *b* 1, 4, 9, 16, ...   *c* 0, 3, 8, 15, ...

6   A firm sends out circulars to 10 500 customers. The circulars cost £2·12 per 1000, the envelopes £4·24 and the postage $2\frac{1}{2}$p each. Find the total cost.

7   If a wavelength is 0·000018 cm, how many wavelengths are there in 1 metre? Express your answer in standard form, $a \times 10^n$, with a rounded off to 2 significant figures.

8   The population of Glasgow was 1101000 in 1939, and 893000 in 1972. Express the decrease as a percentage of the 1939 figure, to 2 significant figures.

9   Calculate the following in the binary scale; check by changing to decimal form.

   a $1101 + 110 - 101$ *b* 10111×11

10   A firm has an overdraft at the bank of £1200 in January, £1560 in February and £1800 in March. What interest does the bank charge each month at the rate of $8\frac{1}{2}\%$ per annum?

11   A dealer bought 120 carpets which had been damaged by fire, and paid an average of £12·50 each for them. 15 were too bad to sell, but he sold the rest at a price which gave him 40% profit on his total outlay. What was his average selling price per carpet?

12   If 1 metre of cable costs 34p, find the cost of 1 km 325 m of cable.

13   A car has length 3·75 m and height 1·5 m. Find the height of a scale model whose length is 9·5 cm.

14   In a class of 30 pupils, 15 like both pop and classical music, 8 like pop only, 5 like classical music only and the rest like neither. What is the probability that a pupil selected at random likes:

   a   both pop and classical music   *b*   classical only   *c*   neither?

15   Express $3540 \times 0·125$ in the form $a \times 10^n$, with a rounded off to 1 decimal place.

16   Use logarithms or slide rule to calculate:

   a $\dfrac{27·8 \times 0·65}{1·23}$ *b* $\sqrt[3]{0·234}$ *c* $\left(\dfrac{4·56}{202 \times 0·067}\right)^2$

   d   the volume of a cylinder with radius of base 12·8 cm and height 48·7 cm. (Take 3·14 as an approximation for π; $V = \pi r^2 h$.)

17   A cuboid is 101_{two} cm long, 11_{two} cm high and 110_{two} cm broad. Calculate the area and volume of the cuboid, using the binary scale. Express the area and volume in decimal form also.

18 After 16 completed innings a batsman's average is 17·25 runs per innings. How many runs must he make in his next (completed) innings to raise his average to 20?

19 A map of the South of France is drawn to the scale 1:250000. The distance between Nice and Menton on the map is 8·4 cm. What is the actual distance? The distance from Avignon to Fréjus is 196·5 km. What length on the map will represent this distance?

20 My annual petrol bill was £67·50 when my car had a consumption of 12 litres per 100 km. What would the bill be after an overhaul which improved the consumption to 10 litres per 100 km, assuming that the same distance was travelled?

21 A book is 5 cm thick. Its covers are each 1·5 mm thick, and it has 1238 pages. Calculate the thickness of each sheet of paper to 0·001 mm. (Both sides of a sheet are numbered pages.)

22 Find: a the length of the side of a square of area 17·5 cm^2
b the radius of a circle of area 70 m^2.

23 From a rectangular sheet of tin 13·5 cm long and 8·4 cm broad two holes are cut, one a square of side 5·3 cm and the other a circle of diameter 7 cm. What area of tin remains?

24 A sportsfield consists of a rectangle 110 m long and 80 m broad, with semicircles outwards on the shorter sides. Calculate the area and perimeter of the field.

25 A theatre occupies a circular plot of ground of diameter 48 m. A path 4 m wide surrounds the building. Calculate the area of the path.

26 Taking the radius of the earth to be $6·4 \times 10^3$ km, calculate its circumference in km and its volume in km^3, expressing your answers in the form $a \times 10^n$ to 2 significant figures. (Take 3·14 for π, and $V = \frac{4}{3}\pi r^3$ for the volume of a sphere.)

27 Calculate the height of a cylindrical tank of radius 65 cm which holds 1700 litres of oil. ($V = \pi r^2 h$ for a cylinder.)

28 A rectangular block of metal is 1·5 m long, and its cross-section is a square of side 8 cm. A cylindrical hole of diameter 7 cm is bored through the block, with its axis parallel to the length of the block. Find, to the nearest kg, the mass of metal left in the block, given that 1 cm^3 weighs 6 g.

29   A business man going to New York on a weekend trip changed £150 to dollars at the rate of 2·36 dollars to the £. How many dollars did he get?

He spent 300 dollars, and changed the rest back to pounds at 2·40 dollars to the £. How much British currency did he get?

30   The world population in 1965 was 3250 million, and was increasing at the rate of 2% per year. Find the population at the end of 1966 and 1967, giving your answers in standard form to 3 significant figures.

31   On a car journey I travelled for 96 km at an average speed of 72 km/h, and then drove for 1 hour 40 minutes at an average speed of 63 km/h. By finding the total distance and the total time, calculate the average speed for the whole journey.

32   A man borrowed £1200. At the end of one year he repaid £600, part of which paid for interest at $9\frac{1}{2}\%$ per annum on the loan and the rest reduced the loan. How much had he to pay at the end of the second year to clear the loan and interest due?

33   Calculate the compound interest on £480 at 3% per annum for 3 years. Interest is added annually, and is calculated on complete pounds of principal.

34   A man has £3000 to invest. Which of the following will give him a greater annual return, and by how much?

   a   Investing the money in a Building Society which pays interest of $4\frac{3}{4}\%$ per annum, tax free.

   b   Buying shares costing 150p each which pay a dividend of 10·05p per share, less income tax at 30p in the £.

35   Use mathematical tables or slide rule to calculate:

   a $\dfrac{35\cdot6 \times 0\cdot316}{8\cdot34}$ b $\sqrt{\dfrac{0\cdot473}{0\cdot734}}$ c $0\cdot631^5$

   d $\dfrac{7\cdot32}{0\cdot056 \times 1\cdot23}$ e $\sqrt[3]{(39\cdot6 \times 2\cdot87)}$ f $\sqrt[3]{\dfrac{0\cdot809}{8900}}$

   g   the length of the diagonal of a rectangle with sides 15 m and 19 m long

   h   the height of a cuboid of length 16·3 mm, breadth 8·31 mm and volume 234 mm^3.

36 A house is valued at £7800 and its contents at £2200. Comprehensive insurance for the house costs $12\frac{1}{2}$p per £100, and for the contents costs 24p per £100. Find the total insurance premium.

37 The rateable value of a house is £98, and the rates are 85p in the £. How much has to be paid in rates?

   If the rates are increased to 97p in the £, how much more has to be paid?

38 A man's annual salary is £4200. His income tax allowances are: personal, £775; children's, £235 and £265; life assurance, $\frac{1}{2}$ premiums of £250.

 a How much income tax does he pay annually at a rate of 30% of his taxable income?

   Monthly deductions are made from the man's salary as follows: income tax; superannuation, £23; graduated pension £4·72; national insurance, £4

 b Calculate his net monthly salary.

39 In a series of class tests the frequency of the scores out of 10 marks was:

Score	0	1	2	3	4	5	6	7	8	9	10
Frequency	0	2	4	7	6	12	22	30	25	15	7

 a Draw a histogram, and calculate the mean score.
 b What is the modal score?
 c Draw a cumulative-frequency curve, and estimate the median and semi-interquartile range of the distribution of marks.

40 Two packs of cards each consist of an ace, king, queen, jack and ten of one suit only. Write out as an array the set of all possible ways of selecting a pair of cards, one from each pack.

   A 'picture card' is a king, queen or jack. Find the probability of selecting:

 a two picture cards b exactly one picture card c at least one picture card.

   Use your answer to c to find the probability of not choosing a picture card.

41   A cube of side 3 cm, made of white wood, has all its faces painted black. If it is now cut up into 1-cm cubes, how many of these will have 0, 1, 2 and 3 black faces? Repeat for a 4-cm cube, cut into 1-cm cubes.

42   The English mathematician, De Morgan, who lived between 1800 and 1900 once said 'I was x years old in the year x^2.' When was he born?

Trigonometry

Triangle Formulae

1 The Cartesian and polar coordinates of a point

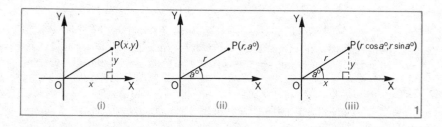

(i)   (ii)   (iii)

The position of a point in the XOY plane can be defined in terms of its Cartesian coordinates (x, y) or its polar coordinates $(r, a°)$.

These coordinates can be related as follows:

$$\cos a° = \frac{x}{r} \qquad\qquad \sin a° = \frac{y}{r}$$

$$\Leftrightarrow \quad x = r\cos a° \qquad\qquad \Leftrightarrow \quad y = r\sin a°$$

So P is the point (x, y), i.e. $(r\cos a°, r\sin a°)$, or $(r, a°)$

Example 1. If the polar coordinates of P are $(6, 40°)$, find its Cartesian coordinates.

$x = r\cos a° = 6\cos 40° = 6 \times 0.766 = 4.60$, to 2 decimal places.
$y = r\sin a° = 6\sin 40° = 6 \times 0.643 = 3.86$, to 2 decimal places.
So P is the point $(4.6, 3.9)$

237

Reminder about the trigonometrical functions of obtuse angles

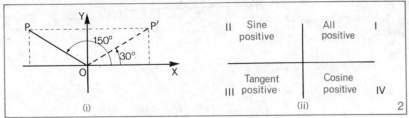

(i)

(ii)

2

For the sine and cosine of 150° the related angle is 30°. By comparing the x and y-coordinates of P and P′, or from Figure 2 (ii), we have sin 150° = sin 30°, and cos 150° = − cos 30°.

In general, the sine of an angle = the sine of its supplement

the cosine of an angle = − the cosine of its supplement.

Example 2. On a radar screen an object is sighted at point P with polar coordinates (5, 152°), as shown in Figure 3. What would be the coordinates of P on the coloured Cartesian grid placed over the radar screen?

$$x = r \cos a° = 5 \cos 152° = -5 \cos 28° = -5 \times 0·883 = -4·42$$
$$y = r \sin a° = 5 \sin 152° = 5 \sin 28° = 5 \times 0·469 = 2·34$$

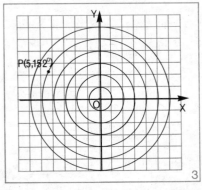

3

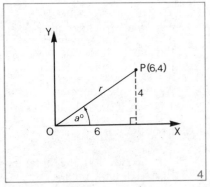

4

Example 3. Find the polar coordinates of the point P(6, 4).
From Figure 4,

$$r^2 = 6^2 + 4^2 = 36 + 16 = 52. \quad \text{So } r = \sqrt{52} = 7·21.$$

$$\tan a° = \frac{y}{x} = \frac{4}{6} = 0·667. \quad \text{So } a° = 33·7°.$$

P has polar coordinates (7·21, 33·7°).

Exercise 1

1 Find (to 2 decimal places) the Cartesian coordinates of the points with polar coordinates:

  a $(2, 20°)$ *b* $(5, 60°)$ *c* $(1, 49°)$

2 Use tables to find the values of:

  a $\sin 150°$ *b* $\sin 100°$ *c* $\sin 175°$

  d $\cos 150°$ *e* $\cos 100°$ *f* $\cos 175°$

3 Find the Cartesian coordinates of the points with polar coordinates:

  a $(2, 110°)$ *b* $(4, 145°)$ *c* $(1, 99°)$

4 Write down the Cartesian coordinates of B in each part of Figure 5.

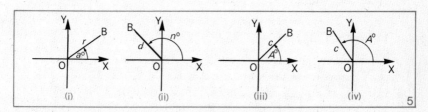

(i) (ii) (iii) (iv) 5

5 The jib of a crane is 10 m long. It is hinged at ground level at one end, and is inclined at 65° to the ground. Calculate:

  a the height of the top of the jib above the ground
  b the distance of the top of the jib from the vertical through the hinge.

6 A ship sails from port on a bearing of 052°, at an average speed of 10 km/h. Calculate its distance after 3 hours:

  a from the port *b* east of the port *c* north of the port.

7 The hour hand of a clock is 6 cm long. Calculate the distance of the tip of the hand:

  a to the right or left of the vertical through the centre of the clock
  b above or below the horizontal through the centre of the clock
    at *(1)* 02 00 *(2)* 16 00 *(3)* 07 30.

8 Find the polar coordinates of the following points:

  a $(4, 3)$ *b* $(5, 12)$ *c* $(3, 1)$ *d* $(-2, 2)$ *e* $(-6, 8)$

2 The sine rule

Introductory examples

1   In Figure 6 (i), $\angle A = 20°$, $\angle C = 90°$, $c = 10$.

  a Are the data sufficient to enable you to *construct* the triangle?
  b Calculate a and b. (a, b, c represent the lengths of the sides opposite angles A, B, C respectively.)

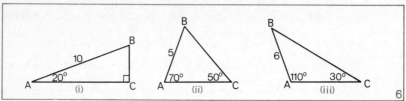

2   In Figure 6 (ii), $\angle A = 70°$, $\angle C = 50°$, $c = 5$.

  a Are the data sufficient to enable you to *construct* the triangle?
  b Why can you not calculate a and b as in question *1*?
  c Introduce right angles by drawing the altitude BP. Can you now see how to calculate BP, BC, AP, PC, AC?

3   In Figure 6 (iii), $\angle A = 110°$, $\angle C = 30°$, $c \doteq 6$.
      Suggest a construction that would enable you to calculate a and b.

General statement and proof of the Sine Rule

In every $\triangle ABC$, $\dfrac{a}{\sin A} = \dfrac{b}{\sin B} = \dfrac{c}{\sin C}$.

$\triangle ABC$ can be placed with its base AC on the *x*-axis, and A or C at the origin as shown in Figures 7 (i) and (ii).

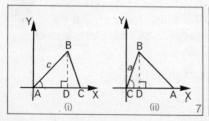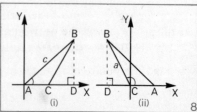

In (i), the y-coordinate of B is $c \sin A$.

In (ii), the y-coordinate of B is $a \sin C$.

It follows that $c \sin A = a \sin C$,

and $\dfrac{a}{\sin A} = \dfrac{c}{\sin C}$ (*dividing each side by sin A sin C*)

With AB placed on OX we obtain

$$\frac{a}{\sin A} = \frac{b}{\sin B}$$

The proof is also true for a triangle with an obtuse angle, as shown in Figure 8.

So we have the Sine Rule for every $\triangle ABC$,

$$\frac{a}{\sin A} = \frac{b}{\sin B} = \frac{c}{\sin C}.$$

Example 1. In $\triangle ABC$, $b = 4 \cdot 2$, $\angle A = 62°$, $\angle B = 46°$. Calculate a.

Use $\dfrac{a}{\sin A} = \dfrac{b}{\sin B} = \dfrac{c}{\sin C}$

and tick or underline the known parts.

$\dfrac{a}{\sin 62°} = \dfrac{4 \cdot 2}{\sin 46°}$

$\Leftrightarrow a \sin 46° = 4 \cdot 2 \sin 62°$ (*by cross-multiplication*)

$\Leftrightarrow \quad a = \dfrac{4 \cdot 2 \sin 62°}{\sin 46°}$

$\qquad = 5 \cdot 2$

Nos	Logs
4·2	0·623
sin 62°	$\bar{1}$·946
	0·569
sin 46°	$\bar{1}$·857
5·15	0·712

Note

(i) The logarithms of the sines are obtained directly from the tables of 'logarithms of sines'.

(ii) If you have a slide rule with trigonometrical scales, the 'proportion' calculation can be done with a single setting of the slide.

(iii) If one angle is obtuse, we use the fact that 'the sine of an angle = the sine of its supplement'.

Example 2. In $\triangle ABC$, $c = 5\cdot8$, $b = 6\cdot7$, $\angle B = 48°$. Calculate $\angle C$.

Use $\dfrac{a}{\sin A} = \dfrac{b\!\!\!/}{\sin B\!\!\!/} = \dfrac{c\!\!\!/}{\sin C}$

$$\frac{6\cdot7}{\sin 48°} = \frac{5\cdot8}{\sin C}$$

$$\Leftrightarrow 6\cdot7 \sin C = 5\cdot8 \sin 48°$$

$$\Leftrightarrow \quad \sin C = \frac{5\cdot8 \sin 48°}{6\cdot7}$$

$$= 0\cdot64 \text{ (by slide rule or logarithms)}$$

Hence $\angle C = 40°$

Exercise 2

1   In $\triangle ABC$, $a = 8$, $\angle A = 49°$, $\angle B = 57°$. Calculate b.

2   In $\triangle ABC$, $b = 2\cdot8$, $\angle A = 53°$, $\angle B = 61°$. Calculate a.

3   In $\triangle ABC$, $a = 25$, $\angle A = 100°$, $\angle C = 28°$. Calculate c.

4   Solve *Introductory Example 2* (p. 240), i.e. calculate a and b in $\triangle ABC$ in which $\angle A = 70°$, $\angle C = 50°$, $c = 5$.

5   Solve *Introductory Example 3*, i.e. calculate a and b in $\triangle ABC$ in which $\angle A = 110°$, $\angle C = 30°$, $c = 6$.

6   In $\triangle ABC$, $a = 17$, $b = 9$, $\angle A = 55°$. Calculate $\angle B$.

7   In $\triangle ABC$, $b = 16$, $c = 21$, $\angle B = 42°$. Calculate $\angle C$.

8   In $\triangle PQR$, $\angle P = 100°$, $PR = 4\cdot5$ cm, $QR = 7\cdot9$ cm. Calculate:

    a $\angle Q$ *b* $\angle R$ (Remember that the sum of the angles of a triangle is 180°.)   *c* PQ

9   In $\triangle XYZ$, $\angle Y = 66\cdot6°$, $\angle Z = 53\cdot2°$, $YZ = 5\cdot15$. Calculate XZ.

10 In $\triangle DEF$, $DE = 123$, $EF = 88$, $\angle F = 114°$. Calculate $\angle E$.

11 A surveyor measures a base line AB 440 m long. He takes bearings of a landmark C from A and B, and finds that $\angle BAC = 48°$ and $\angle ABC = 75°$. Calculate the distances of C from A and from B.

12   D and E are landmarks, E being 15 km east of D. A third landmark F bears 126° from D, and 226° from E. Calculate the distances of F from D and E.

13   Two ships P and Q are 8 km apart. From P, Q bears 100° and another ship R bears 160°. From Q, R bears 200°. Calculate the distances of R from P and from Q.

14   In the metal framework shown in Figure 9 the units are metres. Calculate all the lengths and angles in the triangular frame BCD.

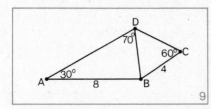
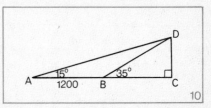

15   From a point A, level with the foot of a hill, the angle of elevation of the top of the hill is 15°. From a point B, 1200 m nearer the foot of the hill, the angle of elevation of the top is 35°. Calculate the height of the hill. (See Figure 10; find BD first.)

---

3 Formulae connecting sinA, cosA and tanA

(i)   In Figure 11, for all positions of P,

$$x^2 + y^2 = r^2 \text{ (Pythagoras' theorem)}$$
$$\Leftrightarrow \quad (r\cos a°)^2 + (r\sin a°)^2 = r^2$$
$$\Leftrightarrow \quad r^2\cos^2 a° + r^2\sin^2 a° = r^2$$
$$\Leftrightarrow \quad \cos^2 a° + \sin^2 a° = 1 \; (\textit{dividing each side by } r^2)$$

Also, $\cos^2 a° = 1 - \sin^2 a°$, and $\sin^2 a° = 1 - \cos^2 a°$

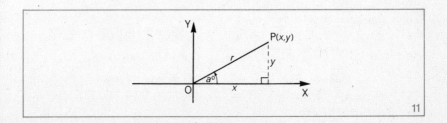

(ii) $\tan a° = \dfrac{y}{x} = \dfrac{r\sin a°}{r\cos a°} = \dfrac{\sin a°}{\cos a°}$

So the following formulae are true for every angle A:

$\cos^2 A + \sin^2 A = 1$

$\cos^2 A = 1 - \sin^2 A$ $\qquad\qquad \tan A = \dfrac{\sin A}{\cos A}$

$\sin^2 A = 1 - \cos^2 A$

Example 1. Verify by using the values of $\cos 30°$, $\sin 30°$ and $\tan 30°$ that:

a $\cos^2 30° + \sin^2 30° = 1$ *b* $\tan 30° = \dfrac{\sin 30°}{\cos 30°}$

From Figure 12, $\cos 30° = \dfrac{\sqrt{3}}{2}$

$\sin 30° = \dfrac{1}{2}$

$\tan 30° = \dfrac{1}{\sqrt{3}}$

a $\quad \cos^2 30° + \sin^2 30°$ $\qquad$ *b* $\quad \dfrac{\sin 30°}{\cos 30°}$

$= (\dfrac{\sqrt{3}}{2})^2 + (\dfrac{1}{2})^2 \qquad\qquad = \dfrac{\frac{1}{2}}{\frac{\sqrt{3}}{2}}$

$= \dfrac{3}{4} + \dfrac{1}{4} \qquad\qquad\qquad = \dfrac{1}{\sqrt{3}}$

$= 1 \qquad\qquad\qquad\qquad = \tan 30°$

Example 2. Given $\sin A = \dfrac{5}{13}$, and $90° < A < 180°$ (i.e. $\angle A$ obtuse), find the values of $\cos A$ and $\tan A$.

$\cos^2 A = 1 - \sin^2 A \qquad\qquad \tan A = \dfrac{\sin A}{\cos A}$

$= 1 - (\dfrac{5}{13})^2 \qquad\qquad\qquad = \dfrac{\frac{5}{13}}{\frac{-12}{13}}$

$= 1 - \dfrac{25}{169} \qquad\qquad\qquad = -\dfrac{5}{12}$

$= \dfrac{169 - 25}{169}$

$= \dfrac{144}{169}$

So $\cos A = \dfrac{12}{13}$ or $-\dfrac{12}{13}$

Since $\angle A$ is obtuse, $\cos A = -\dfrac{12}{13}$

Exercise 3

1   Using a right-angled triangle with two angles of 45°, and sides of length 1, 1 and $\sqrt{2}$ units, show that:

   a $\cos^2 45° + \sin^2 45° = 1$ *b* $\dfrac{\sin 45°}{\cos 45°} = \tan 45°$.

2   Using Figure 12, show that

   a $\cos^2 60° + \sin^2 60° = 1$ *b* $\dfrac{\sin 60°}{\cos 60°} = \tan 60°$

3   Verify that

$$\cos^2 A + \sin^2 A = 1 \quad \text{and} \quad \frac{\sin A}{\cos A} = \tan A \quad \text{when:}$$

   a $A = 135°$ *b* $A = 120°$

4   Sketch the graphs of the sine function and the cosine function for angles from 0° to 360°. With the aid of the graphs show that:

   a $\cos^2 0° + \sin^2 0° = 1$ *b* $\cos^2 90° + \sin^2 90° = 1$
   c $\cos^2 180° + \sin^2 180° = 1$

5   Given $\sin A = \frac{3}{5}$, and $0° < A < 90°$ (i.e. $\angle A$ acute), find the values of $\cos A$ and $\tan A$ by using the formulae

$$\cos^2 A = 1 - \sin^2 A \quad \text{and} \quad \tan A = \frac{\sin A}{\cos A}.$$

6   Given $\sin A = \frac{4}{5}$, and $90° < A < 180°$ (i.e. $\angle A$ obtuse), find the values of $\cos A$ and $\tan A$.

7   Given $\cos A = -\frac{5}{13}$, and $90° < A < 180°$, find the values of $\sin A$ and $\tan A$.

8   Given $\sin A = 0.5$, and $\angle A$ acute, find $\cos A$:
   a   by using the formula   *b*   by first finding A from tables.

9   Given $\cos A = -0.5$, and $\angle A$ obtuse, find $\sin A$:
   a   by using the formula   *b*   by first finding A from tables.

10   Given $\cos A = \frac{4}{5}$ and $\sin B = \frac{5}{13}$, where A and B are acute angles, find the values of:

   a   (*1*) $\sin A$ (*2*) $\tan A$ (*3*) $\cos B$ (*4*) $\tan B$
   b   (*1*) $\sin A \cos B + \cos A \sin B$ (*2*) $\dfrac{\tan A - \tan B}{1 + \tan A \tan B}$

Identities

We can use the formulae

$$\cos^2 A + \sin^2 A = 1, \quad \cos^2 A = 1 - \sin^2 A,$$

$$\sin^2 A = 1 - \cos^2 A \quad \text{and} \quad \tan A = \frac{\sin A}{\cos A}$$

to prove a number of trigonometrical identities.

Examples. Prove that:

1 $(\cos A + \sin A)^2 - 2\cos A \sin A = 1$

2 $\dfrac{\cos A}{\sin A} - \dfrac{\sin A}{\cos A} = \dfrac{2\cos^2 A - 1}{\sin A \cos A}$

Left side

$= (\cos A + \sin A)^2 - 2\cos A \sin A$

$= \cos^2 A + 2\cos A \sin A + \sin^2 A$
$\qquad - 2\cos A \sin A$

$= \cos^2 A + \sin^2 A$

$= 1$

$= right\ side$

Left side

$= \dfrac{\cos A}{\sin A} - \dfrac{\sin A}{\cos A}$

$= \dfrac{\cos^2 A - \sin^2 A}{\sin A \cos A}$

$= \dfrac{\cos^2 A - (1 - \cos^2 A)}{\sin A \cos A}$

$= \dfrac{2\cos^2 A - 1}{\sin A \cos A}$

$= right\ side$

Exercise 3B

Prove the trigonometrical identities in questions *1–11*.

1 $3\cos^2 A + 3\sin^2 A = 3.$

2 $5\cos^2 A = 5 - 5\sin^2 A.$

3 $2\cos^2 A - 1 = 1 - 2\sin^2 A.$

4 $3\cos^2 A - 2 = 1 - 3\sin^2 A.$

5 $(\cos A + \sin A)^2 = 1 + 2\sin A \cos A.$

6 $(\cos A - \sin A)^2 + 2\cos A \sin A = 1.$

7    $(\cos A + \sin A)(\cos A - \sin A) = 2\cos^2 A - 1.$

8    $(\cos A - \sin A)(\cos A + \sin A) = 1 - 2\sin^2 A.$

9    $\cos A \tan A = \sin A.$

10    $\dfrac{1 - \cos^2 A}{\cos^2 A} = \tan^2 A.$

11    $\dfrac{\sin A}{\cos A} + \dfrac{\cos A}{\sin A} = \dfrac{1}{\cos A \sin A}.$

12   State which of the following are true and which are false:

a $\cos^2 B + \sin^2 B = 1$ b $\cos^2 C = 1 + \sin^2 C$

c $\sin^2 A + \cos^2 A = 1$ d $\cos A + \sin A = 1$

e $\cos^2 D - \sin^2 D = 1$ f $\sin^2 X = 1 - \cos^2 X$

g $\dfrac{\cos P}{\sin P} = \tan P$ h $\dfrac{\sin A}{\cos A} = \tan A$

i $\sin A = \cos A \tan A$

4 The cosine rule

Introductory examples

1   In Figure 13 (i), $\angle C = 90°$, $b = 8$, $a = 6$.

a   Are the data sufficient for you to *construct* the triangle?

b   Calculate c, $\angle A$ and $\angle B$.

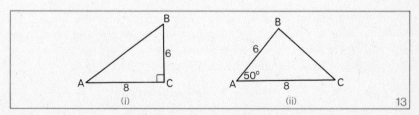

2   In Figure 13 (ii), $\angle A = 50°$, $b = 8$, $c = 6$.

a   Are the data sufficient for you to construct the triangle?

b Why can you not calculate the other sides and angles as in question *1*?

c Write down the Sine Rule for this triangle. Can you now 'solve' the triangle?

We require a formula which can be used when we know the lengths of two sides of a triangle and also the size of the 'included' angle.

General statement and proof of the Cosine Rule

In every $\triangle$ABC, $a^2 = b^2 + c^2 - 2bc \cos$ A.

$\triangle$ABC is placed with its base AC on the x-axis, and A at the origin, as shown in Figures 14 (i) and (ii).

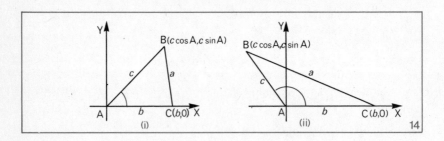

Applying the *distance formula*, $d^2 = (x_1 - x_2)^2 + (y_1 - y_2)^2$ to the points B($c \cos$ A, $c \sin$ A) and C(b, 0), we obtain

$$a^2 = (c \cos A - b)^2 + (c \sin A - 0)^2$$

$$= c^2 \cos^2 A - 2bc \cos A + b^2 + c^2 \sin^2 A$$

$$= b^2 + c^2(\cos^2 A + \sin^2 A) - 2bc \cos A$$

$$= b^2 + c^2 - 2bc \cos A.$$

Note the a^2 on the left side and the angle A on the right side. If vertex B or vertex C were placed at the origin, we would obtain the other formulae included below.

So we have the Cosine Rule for every $\triangle$ABC,

$$a^2 = b^2 + c^2 - 2bc \cos A$$

$$b^2 = c^2 + a^2 - 2ca \cos B$$

$$c^2 = a^2 + b^2 - 2ab \cos C$$

Example. Calculate c in $\triangle ABC$ if $a = 4\cdot36$, $b = 3\cdot84$, $\angle C = 101°$.

	Nos	Logs
	8·72	0·941
	3·84	0·584
	cos 79°	$\bar{1}$·281
	6·40	0·806

$c^2 = a^2 + b^2 - 2ab\cos C$

$\quad = 4\cdot36^2 + 3\cdot84^2 - (2 \times 4\cdot36 \times 3\cdot84 \times \cos 101°)$

$\quad = 19\cdot01 + 14\cdot75 + (8\cdot72 \times 3\cdot84 \times \cos 79°)$

$\quad = 33\cdot76 + 6\cdot40$

$\quad = 40\cdot16$

$c = 6\cdot34$

Note

(i) $\cos 101° = -\cos 79°$ (the cosine of an angle $= -$ the cosine of its supplement)

(ii) Tables of squares and square roots, logarithms and logarithms of cosines, and/or slide rules, are all useful in such calculations.

Exercise 4

Find the length of the third side of each of the triangles in questions *1–6*.

1 $\triangle ABC$, in which $b = 2$, $c = 5$, $\angle A = 60°$.

2 $\triangle ABC$, in which $a = 2$, $b = 5$, $\angle C = 65°$.

3 $\triangle ABC$, in which $a = 2$, $c = 5$, $\angle B = 115°$.

4 $\triangle ABC$ in *Introductory Example 2*, in which $\angle A = 50°$, $b = 6$, $c = 8$.

5 $\triangle PQR$, in which $PR = 4$, $PQ = 3$, $\angle P = 18°$.

6 $\triangle XYZ$, in which $YZ = 25$, $XZ = 30$, $\angle Z = 162°$.

7 In Figure 15, b and c have the same lengths in each triangle, and $\angle A$ is right in (i), acute in (ii), obtuse in (iii).

a Apply the cosine rule to find a^2 in (i). What do you notice?

b Apply the cosine rule to find a^2 in (ii). Which of the following is true?

$\quad\quad$ (*1*) $a^2 = b^2 + c^2$ $\quad$ (*2*) $a^2 > b^2 + c^2$ $\quad$ (*3*) $a^2 < b^2 + c^2$

c Apply the cosine rule to find a^2 in (iii). What is the sign of $\cos A$ in this case? Which sentence in *b* is true?

d Examine the form taken by the cosine rule when A is $0°$, and when A is $180°$. Relate your answers to sketches of 'triangle' ABC.

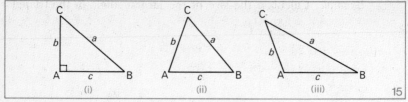

(i)   (ii)   (iii)   15

Questions **8–13** refer to a triangle ABC.

8    $a = 7$, $b = 4$, $\angle C = 53°$. Calculate c.

9    $b = 4·2$, $c = 6·5$, $\angle A = 24°$. Calculate a.

10   $a = 1·64$, $c = 1·64$, $\angle B = 110°$. Calculate b.

11   $a = 18·5$, $b = 22·6$, $\angle C = 72·3°$. Calculate c.

12   $a = 100$, $b = 120$, $\angle C = 15°$. Calculate c.

13   $b = 80$, $c = 100$, $\angle A = 123°$. Calculate a.

14   A town B is 20 km north of town A, and a town C is 15 km north-west of A. Calculate the distance between B and C.

15   Two ships leave port together. One sails on a course of 045° at 9 km/h, and the other on a course of 090° at 12 km/h. How far apart will they be after $2\frac{1}{2}$ hours?

16   Two points A and B on the radar screen shown in Figure 3 (page 238) have polar coordinates $(3, 40°)$ and $(5, 123°)$. If the unit of distance is the kilometre, calculate the actual distance between A and B.

17   Repeat question **16** for two points $P(5, 205°)$ and $Q(6, 350°)$.

To calculate the size of an angle in a triangle,
being given the lengths of the three sides

Given three sides we can construct the triangle, so we should be able to calculate the angles.

$$a^2 = b^2 + c^2 - 2bc \cos A$$

$$\Leftrightarrow 2bc \cos A = b^2 + c^2 - a^2$$

$$\Leftrightarrow \quad \cos A = \frac{b^2 + c^2 - a^2}{2bc}$$

This is another form of the Cosine Rule.

Example. Calculate the size of the largest angle in the triangle with sides of lengths 4·6, 3·2 and 2·8.

Let $a = 4·6$, $b = 3·2$, $c = 2·8$. The largest angle is opposite the largest side a.

$$\cos A = \frac{b^2 + c^2 - a^2}{2bc}$$

$$= \frac{3·2^2 + 2·8^2 - 4·6^2}{2 \times 3·2 \times 2·8}$$

$$= \frac{10·24 + 7·84 - 21·16}{17·92} \ (\textit{using tables of squares})$$

$$= \frac{-3·08}{17·92}$$

$$= -0·172 \ (\textit{by slide rule or logarithms})$$

Since the cosine is negative the angle is obtuse. (It cannot exceed 180° in a triangle.)

$$\angle A = 180° - 80·1°$$

$$= 99·9°$$

Exercise 5

1   Write down formulae for $\cos B$ and $\cos C$ in terms of a, b, c.

2   Write down formulae, in terms of the lengths of the sides, for the following:

     a $\cos P$ in $\triangle PQR$ *b* $\cos Z$ in $\triangle XYZ$ *c* $\cos D$ in $\triangle DEF$

In questions *3–9* calculate the angles named:

3   In $\triangle ABC$, $a = 11$, $b = 10$, $c = 8$; $\angle A$.

4   In $\triangle ABC$, $a = 5$, $b = 10$, $c = 8$; $\angle A$.

5   In $\triangle ABC$, $a = 6$, $b = 7$, $c = 5$; $\angle B$.

6   In $\triangle PQR$, $p = 8$, $q = 10$, $r = 15$; $\angle R$.

7   In $\triangle ABC$, $a = 3·4$, $b = 3$, $c = 1·6$; $\angle A$.

8 In $\triangle$PQR, QR = 7·8, PR = 9·7, PQ = 6·9; the smallest angle.

9 In $\triangle$XYZ, YZ = 20, XZ = 25, XY = 40; the largest angle.

10 If any of the following expressions is the formula for the cosine of an angle of $\triangle$UVW, state which angle:

    a $\dfrac{u^2 + v^2 - w^2}{2uv}$ **b** $\dfrac{u^2 + w^2 - v^2}{2wu}$ **c** $\dfrac{v^2 + w^2 - u^2}{2uw}$

11 Investigate what happens when you try to use the cosine rule to find in turn the angles A, B, C, given that $a = 3$, $b = 4$, $c = 8$. Explain.

12 Calculate the sizes of the angles of a parallelogram in which two adjacent sides and the diagonal from their point of intersection measure 4 cm, 9 cm and 10 cm respectively.

5 The area of a triangle

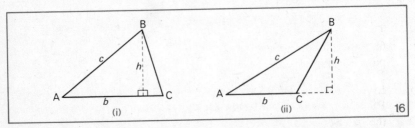

(i)   (ii)   16

Denote the area of triangle ABC by $\triangle$, and the height of triangle ABC from B to AC by h as shown in Figure 16.

Then $\triangle = \frac{1}{2}bh$

In each diagram, $h = c \sin A$.

So $\triangle = \frac{1}{2}bc \sin A$.

Example. Calculate the area of triangle ABC in which $\angle A = 119°$, $b = 2·13$ and $c = 5·46$

$\triangle = \frac{1}{2}bc \sin A$

$ = \frac{1}{2} \times 2·13 \times 5·46 \times \sin 119°$

$ = \frac{1}{2} \times 2·13 \times 5·46 \times \sin 61°$

$ = 5·08$

Nos	Logs
2·13	0·328
2·73	0·436
sin 61°	$\bar{1}$·942
5·08	0·706

Exercise 6

1   Write down two more formulae for the area of triangle ABC, the first in terms of c, a and $\angle B$, the second in terms of a, b and $\angle C$.

2   Calculate the areas of these triangles ABC:

  a $a = 4$, $b = 5$, $\angle C = 30°$. Lengths are in kilometres.
  b $a = 4$, $b = 5$, $\angle C = 150°$. Lengths are in light years.
  c $b = 5$, $c = 4$, $\angle A = 77°$. Lengths are in metres.
  d $c = 10$, $a = 20$, $\angle B = 100°$. Lengths are in kilometres.

3   Calculate the areas of the following triangles in which the lengths are in centimetres:

  a $a = 2·8$, $b = 3·7$, $\angle C = 35·6°$
  b $b = 8·2$, $c = 12·5$, $\angle A = 123°$
  c $a = 3·15$, $b = 2·82$, $\angle C = 32°$
  d $b = 4·93$, $c = 5·39$, $\angle A = 105°$
  e $\angle B = 84·2°$, $a = 8·4$, $c = 21$.

4   Two adjacent sides of a parallelogram have lengths 84 mm and 68 mm, and the angle between them is 72°. Find the area of the parallelogram.

5   In quadrilateral PQRS, $\angle P = 90°$ and $\angle QSR = 54°$. QP = 24 cm, PS = 18 cm and SR = 16 cm. Calculate:
  a   the length of QS   b   the area of PQRS.

6   A regular hexagon ABCDEF is inscribed in a circle, centre O, radius 12 cm long. Calculate the area of:
  a   triangle OAB   b   the hexagon.

7   Calculate the areas of the following regular polygons inscribed in circles:
  a   a pentagon in a circle of radius 10 cm
  b   an octagon in a circle of radius 1 m
  c   an 80-sided polygon in a circle of radius r cm. Using tables, compare this area with the area of the circle.

8   The area of triangle ABC is $20·72$ cm². AB = $6·42$ cm and AC = $8·54$ cm. Find the two possible sizes of angle A.

9   The equal sides of an isosceles triangle are each 4·2 cm long, and the
    area of the triangle is 6 cm². Calculate the two possible lengths of
    the third side.

10 An equilateral triangle of side 2 cm has each side produced 1 cm at
    both ends. Calculate the area of the hexagon formed by joining the
    ends of the lines.

Exercise 7
Miscellaneous questions

1   ABCD is a quadrilateral with angle DAB a right angle. AB = 8 cm,
    BC = 12 cm, CD = 7 cm, DA = 15 cm. Calculate:
    a the length of BD   *b* the size of angle BCD
    c the area of the quadrilateral.

2   Triangle ABC is isosceles, with AB = AC = 12 cm and ∠A = 40°.
    Calculate BC: *a* using the cosine rule *b* using the sine rule
    *c*by drawing altitude AP and using the right-angled triangle ABP.

3   In triangle EFG, FG = 5 cm, ∠EFG = 40°, ∠EGF = 36°. Calcu-
    late the length of EF and the area of the triangle.

4   Calculate all the sides and angles in the metal framework shown in
    Figure 17. The lengths are in metres.

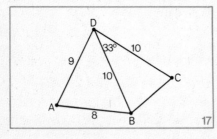 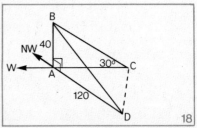

5   Figure 18 shows a vertical mast AB 40 m high, standing on horizon-
    tal ground.
    a When the sun is due west the shadow of the mast is AC, and
    ∠ACB = 30°. Find the length of AC.
    b When the sun is north-west the shadow AD is 120 m long. Find the
    angle of elevation of the sun.
    c Calculate also the distance CD.

6   An aircraft flies from its base 500 km on a bearing 025°, then 750 km on a bearing 280°, then returns to base. Calculate the length and bearing of the last leg of its flight.

7   A coastguard observed a ship 12 km away, due south, and he knew that a lighthouse 10 km away had a bearing 60° east of south. Calculate the distance between the ship and the lighthouse.

If the ship was sailing due east, calculate the shortest distance between it and the lighthouse.

8   Prove that:

a $4\sin^2 A - 3 = 1 - 4\cos^2 A$

b $(1 - \sin^2 A)\tan^2 A = \sin^2 A$

c $(\cos A - \sin A)^2 = 1 - 2\sin A \cos A$

d $\cos^4 A - \sin^4 A = \cos^2 A - \sin^2 A$

Summary

1   *The coordinates of a point* can be given in the forms:
    (i) Cartesian coordinates (x, y), i.e. $(r \cos a°, r \sin a°)$
    (ii) polar coordinates $(r, a°)$.

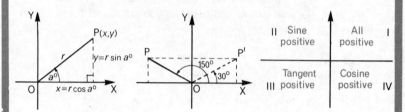

2   The sine of an angle = the sine of its supplement;
    the cosine of an angle = − the cosine of its supplement.

3   The *Sine Rule* for every triangle ABC

$$\frac{a}{\sin A} = \frac{b}{\sin B} = \frac{c}{\sin C}$$

4   *Formulae* connecting sin A, cos A and tan A.
    $\cos^2 A + \sin^2 A = 1$

    $\cos^2 A = 1 - \sin^2 A$ $\tan A = \dfrac{\sin A}{\cos A}$

    $\sin^2 A = 1 - \cos^2 A$

5   The *Cosine Rule* for every triangle ABC

$$a^2 = b^2 + c^2 - 2bc \cos A;$$

$$\cos A = \frac{b^2 + c^2 - a^2}{2bc}$$

6   The *area* of every triangle ABC.

$$\triangle = \tfrac{1}{2}bc \sin A$$

Revision Exercise

Revision Exercise on Chapter 1
Triangle Formulae

Revision Exercise 1

1   Find the Cartesian coordinates of the points with polar coordinates:

   a $(6, 60°)$ *b* $(3, 30°)$ *c* $(7, 90°)$ *d* $(10, 135°)$

2   A is the point $(5, 20°)$ and B is the point $(2, 80°)$. AP and BQ are drawn perpendicular to the *x*-axis, and AM and BN perpendicular to the *y*-axis. Find the Cartesian coordinates of A and B. Hence find the lengths of PQ and MN.

3   Find the polar coordinates $(r, a°)$ of the points:

   a $(7, 0)$ *b* $(0, 3)$ *c* $(-2, 0)$ *d* $(7, 24)$

4   Use tables to verify that $\sin^2 A + \cos^2 A = 1$, and

$$\frac{\sin A}{\cos A} = \tan A \text{ when } A = 86°.$$

5   Given $\sin A = \frac{12}{13}$, use the formulae in question *4* to find the values of $\cos A$ and $\tan A$ in the cases when: *a* $0° < A < 90°$ *b* $90° < A < 180°$.

6   Particulars of ten different triangles are given in the following table.

	a	b	c	A	B	C	Δ
(1)			7	68°	72°	?	
(2)	4	?		50°	70°		
(3)	6	5				48°	?
(4)	5	4	6		?		
(5)		?	4	35°			10
(6)	6	8	?			54°	
(7)	9	6		72°	?		
(8)	10	?	12		110°		
(9)	7	4	5	?			
(10)	4	6		40°	?		

Units of length are centimetres, and units of area are square centimetres. In each case find the element against which a question mark is placed.

7   In triangle ABC, angle $A = 40.3°$, angle $B = 64.5°$, $AC = 8.8$ cm. Calculate the lengths of AB and BC.

8   PQRS is a parallelogram in which $PQ = 7.5$ units, $QR = 9.2$ units, and angle $PQR = 42.6°$. Calculate:
  a the lengths of the diagonals   b   the area of the parallelogram.

9   In quadrilateral ABCD, $AB = 12$ cm, $AC = 15$ cm, $CD = 8.0$ cm, $\angle BAC = 25°$, $\angle DCA = 110°$. Calculate the area of the quadrilateral.

10   A stained-glass window consists of 150 fitting regular hexagonal glass panels. If the side of each panel is 8 cm long, calculate the total glass area.

11   An aeroplane flies from base on course 084° for 280 km, and then on course 021° for 155 km. By using one of the triangle formulae directly, or otherwise, find how far the aeroplane then is from base. Find also the course which the navigator should set for a straight flight home.

12   A hillside descends at a constant slope to the bank of a river. From a point on the hillside, in the direction in which a horizontal bridge is to be built, the angles of depression of the proposed ends of the bridge are 18° and 15°, and the distance along the slope to the nearer of these positions is 960 m. Calculate the length of the bridge.

13   Triangle ABC, in which $BC = 12$ cm, is inscribed in a circle of radius 9 cm. By calculating the angle which BC subtends at the centre, or otherwise, find the two possible sizes of angle BAC.

14   PQ and XY are straight lines which bisect each other at M. $PQ = 30$ cm, $XY = 16$ cm, and $\angle PMY = 120°$. Calculate the area and perimeter of the quadrilateral PXQY.

15   Prove the following trigonometrical identities:
  a $2\cos^2 A + 2\sin^2 A - 2 = 0$ b $(\cos A - \sin A)^2 = 1 - 2\sin A \cos A$

  c $\dfrac{1}{1+\sin A} + \dfrac{1}{1-\sin A} = \dfrac{2}{\cos^2 A}$ d $\cos^2 A(1 + \tan^2 A) = 1$

16   Given that $p - q = \cos A$ and $\sqrt{(2pq)} = \sin A$, prove that $p^2 + q^2 = 1$.

Cumulative Revision Exercise

1   OX is the positive part of the x-axis. If P is the point (8, 6) and Q is (−3, −3), write down the values of the cosine, sine and tangent of angles XOP and XOQ.

2   In Figure 1 the units are cm. Calculate h.

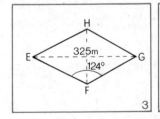
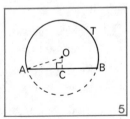

3   In Figure 2, calculate a.

4   From a point P the angle of elevation of a hill top 245 m high is 25°. Calculate the distance from P to the hill top.

5   The structure of a bridge is in the form of a rhombus EFGH as shown in Figure 3. If ∠ EFG = 124° and EG = 325 m, calculate the height of H above F.

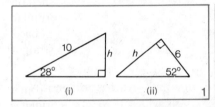
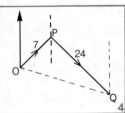
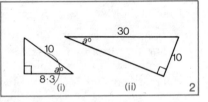

6   A ship sails 7 km from O on a course 045°, as shown in Figure 4. It then sails 24 km on a course 135° to Q.

a Calculate the sizes of angles OPQ and OQP, and the distance OQ.

b By drawing north–south lines at P and Q as shown, and filling in the sizes of angles, calculate the bearing of O from Q.

7 In Figure 5, ACBTA represents the cross-section of an underground tunnel. ATB is an arc of a circle with centre O and radius 7·5 m, and the floor AB is horizontal and 12 m wide. Calculate ∠OAC and the height of the tunnel.

8 In △ABC, ∠A = ∠C = 45°, and AB = BC = 1 cm. *a* Write down: (*1*) the length of AC (*2*) the values of cos 45°, sin 45° and tan 45°. *b* Deduce the values of cos 135°, sin 135° and tan 135°.

9 In △PQR, PQ = QR = RP = 2 cm, and RS is an altitude of the triangle.

a Write down: (*1*) the lengths of PS and SR (*2*) the values of cos 30°, sin 30°, tan 30°, cos 60°, sin 60°, tan 60°.

b Deduce the values of cos 150°, sin 210°, tan 330°, cos 300°, sin 240°, tan 120°.

10 Figure 6 (i) shows a force of 12 newtons represented in magnitude and direction by $\overrightarrow{OB}$. The force has a horizontal component represented by $\overrightarrow{OA}$ and a vertical component represented by $\overrightarrow{OC}$. Calculate the magnitudes of the horizontal and vertical components.

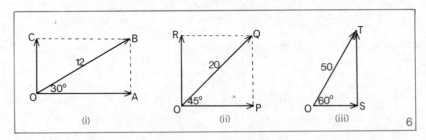

(i) (ii) (iii)

6

11 Repeat question *10* for the force of 20 newtons represented by $\overrightarrow{OQ}$ in Figure 6 (ii).

12 Repeat question *10* for the force of 50 newtons represented by $\overrightarrow{OT}$ in Figure 6 (iii).

13 Find the values of:

a sin 21° *b* sin 121° *c* sin 221° *d* sin 321° *e* sin 421°

14 Which of the following are true and which are false?

a $\tan 30° = \tan 330°$ *b* $\cos 70° = \cos 290°$

c $\sin 400° = -\sin(-40°)$ *d* If $A = 45°$, $\sin A = \cos A$.

e If $\sin B = \cos B$, $B = 45°$ if $0° \leqslant B \leqslant 360°$.

15a Sketch the graph of the sine function from $0°$ to $720°$.

  b For $0 \leqslant x \leqslant 720$, $x \in R$, give the solution sets of
    (1) $\sin x° = 0$ *(2)* $\sin x° = 1$.

16a Sketch the graph of the cosine function from $-360°$ to $+360°$.

  b For $-360 \leqslant x \leqslant +360$, $x \in R$, give the solution sets of
    (1) $\cos x° = 1$ *(2)* $\cos x° = -1$.

17 Using the sine graph, find for what replacement of $2x$, $\sin 2x° = 1$, $0 < x < 180$. Hence write down the replacement for x for which $\sin 2x° = 1$, $0 < x < 180$.

18 Solve $\cos 2x° = -1$, $0 < x < 180$.

19 Copy and complete the table below. Plot the corresponding points, and sketch the graph of the function $f(x) = 2 \sin x°$ for $0 \leqslant x \leqslant 360$, $x \in R$.

x	0	90	180	270	360
$\sin x°$	0	1			
$f(x) = 2\sin x°$	0	2			

(In questions *19*, *20* and *21* assume that each graph consists of a smooth curve.)

20 Repeat question *19* for the graph of the function $g(x) = \sin^2 x°$.

21 Repeat question *19* for the graph of the function $h(x) = \sin x° + \cos x°$.

22 An aircraft flies 246 km on a course $063°$, then 538 km on a course $308°$. Calculate its distance *a* east or west of its starting point *b* north or south of its starting point. (Calculate each separately for each part of the course by drawing right-angled triangles in the diagram.)

23 The polar coordinates of the positions of four ships from a control ship O are $A(2, 50°)$, $B(10, 147°)$, $C(5, 200°)$, $D(10, 355°)$. Express the positions as Cartesian coordinates with respect to O.

24 Express each of the following as trigonometrical functions of acute angles.

a $\sin 100°$ *b* $\cos 96°$ *c* $\sin 149°$ *d* $\cos 170°$ *e* $\sin 485°$

25   In $\triangle ABC$, $a = 29$, $b = 36$, $\angle B = 51°$. Calculate $\angle A$ and $\angle C$.

26   In $\triangle DEF$, $d = 110$, $\angle D = 138°$, $\angle F = 21°$. Calculate e and f.

27   In $\triangle PQR$, $PQ = 8.25$, $\angle Q = 36.5°$, $\angle P = 63.5°$. Calculate QR.

28   In $\triangle STV$, $ST = 36$, $TV = 27$, $\angle STV = 111.5°$. Calculate the area of the triangle.

29   In $\triangle XYZ$, $XY = 3.8$, $\angle X = 63.4°$, $YZ = 3.4$. Calculate the size of the largest angle.

30   Calculate the size of the largest angle in the triangle with sides 5, 6 and 10.

31   In $\triangle ABC$, $b = 2.5$, $c = 3.5$, $\angle A = 130°$. Calculate a.

32   Calculate the area of $\triangle ABC$ in question *31*.

33   A ship sails 80 km on a course 065°, then 100 km on a course 165°. Calculate the distance and bearing from its starting point.

34   Which of the following are true and which are false?

a $\sin^2 A + \cos^2 A = 1$ *b* $\sin^2 \theta + \cos^2 \theta = 1$

c $\sin^2 6X + \cos^2 6X = 1$ *d* $\cos^2 Z = 1 - \sin^2 Z$

e $\tan A = \dfrac{\sin A}{\cos A}$ *f* $\dfrac{\sin 2A}{\cos 2A} = 2 \tan A$

35   Given $\sin A = \frac{3}{5}$, find $\cos A$ and $\tan A$ if: *a* $0° < A < 90°$

                                                                        b $90° < A < 180°$.

36   Prove that the following identities are true:

a $5\sin^2 A + 3\cos^2 A = 2\sin^2 A + 3$

b $4\cos^2 A - 2\sin^2 A = 6\cos^2 A - 2$

c $(2\cos A + 3\sin A)^2 + (3\cos A - 2\sin A)^2 = 13$

d $\tan A + \dfrac{1}{\tan A} = \dfrac{1}{\sin A \cos A}$

e $(\sin \theta + \cos \theta)^2 - (\sin \theta - \cos \theta)^2 = 4\sin \theta \cos \theta$

Computer Studies

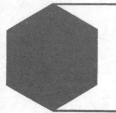

Computers
and their Use in Society

1   The computer era

Nowadays computers affect all our lives, whether we realize it or not. For example, Glasgow installed a large computer in 1966 to assist in the running of the city, replaced it by a much larger one in 1969, and by a still larger one in 1973. With a population of nearly 1 million people, 55 000 of whom are Council employees earning a total of over £80 million each year, massive tasks involving control of manpower and finance have to be faced in such a community.

Glasgow has made use of its computers in many ways, including the preparation of payrolls and accounts; payments to creditors and pensioners; organisation of health records; control of rent accounts and allocation of houses; highway design; road accident records and patterns; research; analysis of each population census to assess and forecast the distribution of population and the associated needs in education, social work and housing; etc.

In this chapter we study some aspects of computers, very briefly, in Sections 2-5, and some common uses of computers in Section 6. The chapter ends with Topics to Explore and Projects to Pursue, so that you can investigate some of the applications of computers which multiply year by year in this computer era.

2 *Electronic computers*

Figure 1 shows a modern computer, or to give it its full title, an *automatic electronic digital computer.**

A *computer* is simply a calculating device; examples include a hand calculating machine, a slide rule and even a clock.

A *digital computer* carries out its calculations by counting. A hand calculating machine is an example of a digital computer.

An *electronic digital computer* is a digital computer in which most of the components work electronically. The transistor is a typical component of this kind. The very high speed of operation of such a computer is mainly due to these electronic parts.

An *automatic electronic digital computer* is one which, when set in motion, will carry out its instructions from beginning to end without any human intervention, that is, automatically. Many households contain an automatic washing machine which can deal with several sets of instructions for washing and rinsing different types

* Photograph supplied by Systemshare Ltd

of materials. Its correct operation depends on the setting of dials, or the insertion of a specially shaped plate into the machine; in either case the housewife can provide the required instructions by selecting the correct *program*. Once the washing machine is set in motion, these instructions are followed automatically.

3 Programs and data

Modern computers have two main functions. One is to store information and the other is to process information or carry out calculations very rapidly. Often these two functions are combined, as for example in the preparation of gas, electricity and telephone accounts, and in the preparation of wage slips and salary cheques. To perform its tasks a computer has to be supplied with:

(i)   the information to be processed, called the *data*

(ii) instructions for the processing, called the *program*.

We saw in Book 5 that to communicate with a computer we have to write the instructions in a language that it can 'understand', and we have to present the instructions in a form that it can 'read'.

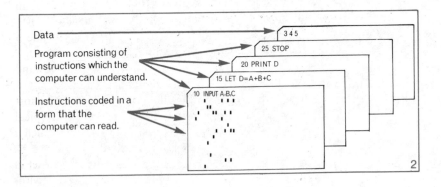

Figure 2 shows a program written in the programming language BASIC which the computer can *understand*, and presented on punched cards in a code that the computer can *read*.

4 The main parts of a computer

An outline of the action in processing information by computer is shown in Figure 3.

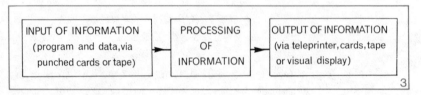

The main components of a computer, which carry out the above process, are:

(i)   The *input device*, which feeds information (programs and data) into the machine.

(ii) The *memory*, which stores information and makes it available when needed.

(iii) The *arithmetic unit*, which performs the calculations.

(iv) The *output device*, which prints out results or may display them on a screen.

(v)   The *control unit*, which enables all the different parts of the computer to function in the correct order.

Figure 4 indicates the arrangement of the various parts of a computer.

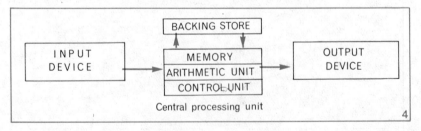

The *central processing unit* consists of the memory, the arithmetic unit and the control unit, and the various devices attached to this unit are called *peripherals*. Extra storage facilities are provided by *backing stores* which can hold vast quantities of information over a long period of time, for example in connection with hire purchase

accounts, income tax, etc. This information may be stored in magnetic tapes, as in a tape recorder, or on magnetic discs rather similar
to the discs used on record players.

Project. Copy a computer

Twelve to eighteen pupils are required, as listed below:

QR, the questioner
PR, the programmer
CW, the card writer
OP, the operator
MS, the messenger
ID, the input device
SA, storage location A
SB, storage location B
SC, storage location C
SD, storage location D
AU, the arithmetic unit
OD, the output device
CU, the control unit

Suppose QR asks 'What is
$368 + 255 - 149$?'.
With the help of the others, PR
sketches a flow chart if necessary,
and writes a suitable program in
BASIC, for example:

10 LET A = 368
20 LET B = 255
30 LET C = 149
40 LET D = A + B − C
50 PRINT D
60 STOP

CW prepares a card for each line
of the program.

Computer operation

Instruction number	from	to	Instruction
1	CU	OP	Take punched cards from CW, and give them to ID.
2	CU	MS	Take *copy* of card 1 from ID, and give it to SA.
3	CU	MS	Take *copy* of card 2 from ID, and give it to SB.
4	CU	MS	Take *copy* of card 3 from ID, and give it to SC.
5	CU	MS	Take *copy* of number from SA, and give it to AU.
6	CU	MS	Take *copy* of number from SB, and give it to AU.
7	CU	AU	Add these numbers, write result, destroy numbers.
8	CU	MS	Take *copy* of number from SC, and give it to AU.
9	CU	AU	Subtract this number from last, write result, destroy numbers.
10	CU	MS	Take result from AU, and give it to SD.
11	CU	MS	Take *copy* of result from SD, and give it to OD.
12	CU	OD	Print out result.
13	CU	MS	Take printout from OD, and give it to QR.

Repeat the project for different kinds of calculation; for example,
calculate:

  a $(4\cdot6 \times 2\cdot5) + 6\cdot8$ *b* $9\cdot5 \times 4\cdot2 \div 15$ *c* $(14\cdot7 - 6\cdot9)^2$

5 Binary notation and machine code

Figure 2 showed a program and data in a form which we said the computer could read and understand. In fact it is not strictly true to say that a computer can understand BASIC in the form in which we write it and present it to the machine. A computer can accept instructions only if they consist of numbers in binary form. The reason is that the logical circuits used in the construction of computers are of a two-state type, that is they are based on true–false, yes–no, on–off type responses. Therefore the information required to operate them can be rapidly transmitted between the various parts of the computer as electrical impulses whose presence or absence can be represented by two symbols. If 1 and 0 are chosen as these symbols, messages in binary form can be readily processed.

For example, the BASIC statement PRINT D (which tells the computer to print out the number in storage location D) is replaced by numerical instructions which are then expressed in binary form.

If PRINT is replaced by the number 14, and D by 29,
PRINT D becomes 14 29, which is then written 1110 11101.

A program written in a 'numeric code' like this is said to be written in *machine code*, which is the only form understood by a computer. In practice, the computer translates instructions from a 'high level' language like BASIC into its own machine code by means of a special program (supplied by the manufacturer) called a *compiler*, as indicated in Figure 5.

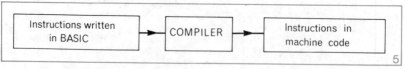

Writing programs in machine code is difficult and tedious, and as each type of computer has its own machine code, it is not possible to run a machine code program which was written for one make on a different make of machine. So there are several great advantages in writing programs in a language like BASIC which uses English words instead of numeric codes, and which can be used with many different types of computer.

6 Some uses of computers in society today

Computers are now being used in a great variety of ways, as indicated in the *Topics to Explore* in Section 7. In the present section we study three common uses of computers in society today.

(*i*) *Payrolls*

Every week millions of people in Britain receive pay-packets and payslips of the kind we saw in Social Arithmetic on page 165. The preparation of these payslips involves a large number of calculations based on certain information about the persons concerned. For example, gross pay and overtime have to be worked out, appropriate deductions made for income tax, national insurance, graduated pension, etc., and the net pay finally calculated.

Several of these items may vary from week to week for every wage-earner, so that in a large firm the preparation of the payslips is a major operation. As there are well-established rules for carrying out all the necessary processes and calculations, much of the work can be done quickly and efficiently by computer. A computerised payroll system might require information similar to that shown in Figure 6.

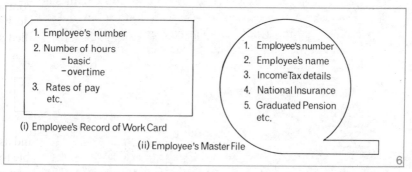

1. Employee's number
2. Number of hours
   - basic
   - overtime
3. Rates of pay
   etc.

(i) Employee's Record of Work Card

1. Employee's number
2. Employee's name
3. Income Tax details
4. National Insurance
5. Graduated Pension
   etc.

(ii) Employee's Master File

6

Each week a new Record of Work Card is made for every employee, and the Employee's Master File (usually in the form of information stored in magnetic tape) is updated, before the payslips are prepared. A suitable computer program is made on the basis of the information extracted; as a result the payslips are produced and the correct amount of money is paid to each employee. At the same time, other useful information, such as the total payroll figures, can be readily printed out for the use of the firm or its accountants.

(ii) Invoices, accounts and credit control

In many buying and selling transactions today no money changes hands at the time, but records are kept by the shop or firm, and invoices and accounts are issued to customers as necessary. An efficient system for recording purchases and payments, and issuing accounts, is essential; one based on computer control and processing is widely used.

Suppose that a customer has an account with a certain shop, and purchases the items shown in Figure 7.

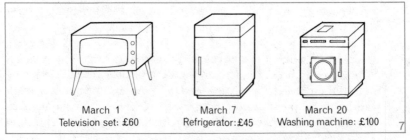

| March 1 | March 7 | March 20 |
| Television set: £60 | Refrigerator: £45 | Washing machine: £100 |

7

As the customer has an account with the shop she does not have to pay cash at the time of purchase. Within a few days of each purchase the shop sends her an *invoice*, giving details of the transaction, and at the end of the month she will receive a *statement* showing cash paid and cash still owing, as shown in Figure 8.

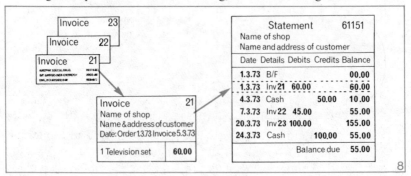

Invoice   23
Invoice   22
Invoice   21

Invoice	21
Name of shop	
Name & address of customer	
Date: Order 1.3.73 Invoice 5.3.73	
1 Television set	60.00

Statement				61151
Name of shop				
Name and address of customer				
Date	Details	Debits	Credits	Balance
1.3.73	B/F			00.00
1.3.73	Inv 21	60.00		60.00
4.3.73	Cash		50.00	10.00
7.3.73	Inv 22	45.00		55.00
20.3.73	Inv 23	100.00		155.00
24.3.73	Cash		100.00	55.00
			Balance due	55.00

8

Most firms give each customer a *credit limit*. When the customer's purchases reach this limit the firm sends a letter requesting the payment of cash before further purchases are made. This process is called *credit control*.

Obviously the whole cycle described above is of a repetitive nature. It involves the storage of a great deal of information and the performing of a large number of calculations, so it is very suitable for a computer-based operation.

In Figure 9, the input to the computer in such a system is shown in the form of 'New Accounts and Alterations' and 'Transactions'. This input updates the data in two files in the computer's memory. These files are called the 'Customer Record File' and the 'Customer Account File'. Using all the information supplied the computer will produce:

(i)   the monthly statement to the customer, shown in Figure 8
(ii) a customer's credit analysis slip for the firm's records.

At the same time, the customer's files will be updated, ready for further purchases.

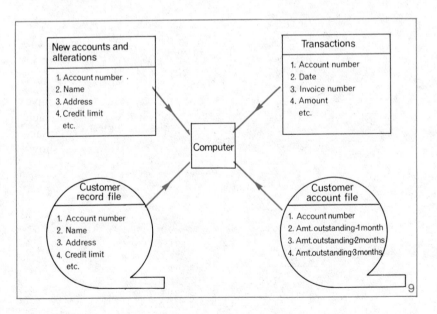

(iii) Preparing an electricity account

How long would you take to calculate the amount due for the following electricity account?

Present reading (S)	= 17 536		*Charge per unit*
Previous reading (R)	= 14 246	(i)	2·9 p per unit for the first 60 units
Number of units used (U) =	3 290	(ii)	0·8 p per unit for remaining units

If you actually try the above calculation you can work out the time you would take to calculate all the accounts for a village with 1000 users of electricity, or for a town with 1 000 000 users.

Such a task is comparatively quickly done by computer. Here is a simplified flow chart and BASIC program to calculate a customer's account.

Let C = customer's number; R = previous reading; S = present reading.

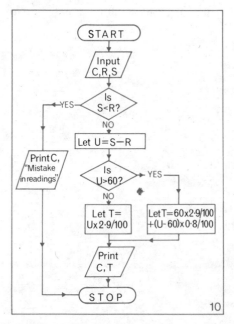

```
10    REM ELECTRICITY
         ACCOUNT
20    INPUT C, R, S
30    IF S < R THEN 110
40    LET U = S − R
50    IF U > 60 THEN 80
60    LET T = U*2·9/100
70    GOTO 90
80    LET T = 60*2·9/100 +
         (U − 60)*0·8/100
90    PRINT C, T
100   GOTO 120
110   PRINT C, "MISTAKE
         IN READINGS"
120   STOP
```

The process of preparing an electricity account is more complicated in practice as a variety of checks are incorporated to avoid errors.

As well as carrying out the calculations, the computer will print out all the information that appears on the account, and may even type out both the account and the envelope addressed to the user.

Exercise

Follow through the flow chart in Figure 10 for the following data; if possible, prepare and run the associated BASIC programs.

| *Present reading* (S) | 700 | 1041 | 3920 | 235 | 12 567 |
| *Previous reading* (R) | 650 | 1041 | 3600 | 352 | 10 818 |

7 Topics and projects

Topics to explore

Choose one or two of the following areas in which computers are having an increasing impact on modern society, and find out more about them.

1 *Science and its applications*
Space travel; weather forecasting; research and design; guidance and control of aircraft, nuclear submarines and missiles.

2 *Business and commerce*
Data processing; payrolls; bank statements; gas, electricity and telephone accounts; airline ticket reservations; National Giro Service; credit cards; stock control; hire purchase.

3 *Industry*
Machine-tool control; production control, as in power stations; export and import control; freight terminal operation.

4 *Record storage and retrieval*
Libraries; medical records and diagnosis; criminal records.

5 *Traffic control*
Linked traffic-light system; road network design.

6 *Education*
Computer assisted instruction (CAI); processing examination marks; applications in certain school subjects, e.g. mathematics, business studies, science, geography.

7   *Miscellaneous*
Translation from one language to another, including code-breaking; strategy in war games, chess matches, etc.; design of architecture; composition of 'musical' scores; investigation of problems in number theory.

Projects to pursue

1   *Prepare a scrapbook* or a classroom wall display of photographs, diagrams and other information about computers from newspapers, brochures, etc.

2   *Follow up a visit* to a computer installation, or the viewing of a film or a television programme about computers, by a discussion or an essay or a study of careers associated with computers.

Answers

Answers

Algebra—Answers to Chapter 1

Page 4 Exercise 1

1	$a+b$	*2*	$x+2y$	*3*	$3p+2q$	*4*	$p+q$	*5* $b+1$

1 $a+b$ *2* $x+2y$ *3* $3p+2q$ *4* $p+q$ *5* $b+1$

6 $y+x$ *7* c^2+1 *8* $x+2y$ *9* $c-d$ *10* $3m-2n$

11 $q-2$ *12* $2-5x$ *13* $3(a+b)$ *14* $8(x+y)$ *15* $4(x-y)$

16 $5(c-d)$ *17* $2(b+2c)$ *18* $3(3m-2n)$ *19* $a(x+y)$ *20* $b(x-y)$

21 $a(x+1)$ *22* $c(x^2+1)$ *23* $a(a+b)$ *24* $p(p-q)$ *25* $x(x+1)$

26 $y(y-1)$ *27* $t^2(t+1)$ *28* $k^2(k-1)$ *29* $q(p+r)$ *30* $d(c-e)$

31 $4(2x+3y)$ *32* $7(5a-2b)$ *33* $2a(a+3b)$ *34* $3a(4c-3b)$ *35* $ab(c+d)$

36 $2n(3n-1)$ *37a* 3400   *b* 10·8   *c* 34   *d* 10

38 $2(a+b+c)$ *39* $a(x+y+z)$ *40* $c(x-y+z)$ *41* $2(2x+y-3z)$

42 $2p(2p-5q)$;   76.

Page 5 Exercise 1B

1 $3(2m+3n)$ *2* $4(3p+2q)$ *3* $x(a-b)$ *4* $4(5-x)$ *5* $x(1-x)$

6 $y(x+1)$ *7* $m(v-u)$ *8* $a(2x+y)$ *9* $2a(3x-2y)$ *10* $3a(3b-4c)$

11 $2x(3x+2)$ *12* $3a(5a-2b)$ *13* $\frac{1}{3}(p-q)$ *14* $\frac{1}{2}h(a+b)$ *15* $t(u+\frac{1}{2}at)$

16 $2pq(5p+4q)$ *17* $\frac{1}{2}gt(T-t)$ *18* $\frac{1}{2}(1+a)$ *19* $\frac{1}{2}(m-3)$ *20* $2\pi r(h+r)$

21 $3(a+2b+3c)$ *22* $2(2x+y+4z)$ *23* $a(b+c-a)$ *24* $3(3p-2q-r)$

25 $x(3x-2y+3)$ *26* $5x(3x+2y-4z)$ *27* F   *28* F   *29* T

30 T   *31* F   *32* T   *33* F   *34* F   *35* T   *36* T

37 $a-(b-c)$ *38* $a-(b+c)$ *39* $a-3(b+2c)$ *40* $a-3(b-2c)$

41 $a(a+x)+y(a+x)$ *42* $a(b-x)-x(b-x)$ *43* $(a+b)(x+y)$

44 $(a+x)(a+b)$ *45* $(a+b)(a-c)$ *46* $(c-a)(x-y)$

47 $(a+b)(a+c)$ *48* $(x-y)(a+b)$ *49* $(a+b)(a-c)$ *50* $(a-b)(a-c)$

Page 6 Exercise 2

1 $(a-b)(a+b)$ *2* $(c-d)(c+d)$ *3* $(p-q)(p+q)$

4 $(x-y)(x+y)$ *5* $(x-2)(x+2)$ *6* $(a-3)(a+3)$

7 $(b-4)(b+4)$ *8* $(c-5)(c+5)$ *9* $(a-1)(a+1)$

10 $(b-2)(b+2)$ *11* $(c-6)(c+6)$ *12* $(d-10)(d+10)$

13 $(1-k)(1+k)$ *14* $(3-m)(3+m)$ *15* $(4-n)(4+n)$

16	$(8-p)(8+p)$	*17*	$(r-s)(r+s)$	*18*	$(u-v)(u+v)$
19	$(m-n)(m+n)$	*20*	$(y-z)(y+z)$	*21*	$(a-2b)(a+2b)$
22	$(c-3d)(c+3d)$	*23*	$(e-7f)(e+7f)$	*24*	$(g-9h)(g+9h)$
25	$(4x-y)(4x+y)$	*26*	$(2p-q)(2p+q)$	*27*	$(5r-s)(5r+s)$
28	$(6u-v)(6u+v)$	*29*	$(2x-3y)(2x+3y)$	*30*	$(3a-4b)(3a+4b)$
31	$(4c-3d)(4c+3d)$	*32*	$(2e-5f)(2e+5f)$	*33*	$2(a-b)(a+b)$
34	$2(a-2b)(a+2b)$	*35*	$2(a-3b)(a+3b)$	*36*	$2(a-5b)(a+5b)$
37	$5(p-q)(p+q)$	*38*	$5(3p-q)(3p+q)$	*39*	$3(p-2q)(p+2q)$
40	$3(3p-q)(3p+q)$	*41*	$9(a-2y)(a+2y)$	*42*	$3(b-3)(b+3)$
43	$4(c-5)(c+5)$	*44*	$5(d-1)(d+1)$	*45*	$8(x-2y)(x+2y)$
46	$2(2x-5y)(2x+5y)$	*47*	$(p-2qr)(p+2qr)$	*48*	$(ab-2c)(ab+2c)$

Page 7 Exercise 2B

1	$(a-1)(a+1)(a^2+1)$	*2*	$(c-2)(c+2)(c^2+4)$	*3*	$(d-3)(d+3)(d^2+9)$
4	$(1-x)(1+x)(1+x^2)$	*5*	$(a-b)(a+b)(a^2+b^2)$	*6*	$(x-y)(x+y)(x^2+y^2)$
7	$(p-q)(p+q)(p^2+q^2)$	*8*	$(1-2c)(1+2c)(1+4c^2)$	*9*	$3(x-2)(x+2)(x^2+4)$
10	$4(y-1)(y+1)(y^2+1)$	*11*	$2(z-3)(z+3)(z^2+9)$	*12*	$5(2-a)(2+a)(4+a^2)$
13	$(a+b-c)(a+b+c)$	*14*	$(x+y-z)(x+y+z)$	*15*	$(p+q-2)(p+q+2)$
16	$(m-n-p)(m-n+p)$	*17*	$(b-c-a)(b-c+a)$	*18*	$(m-n-1)(m-n+1)$
19	$(a-b+c)(a+b-c)$	*20*	$(d-e+f)(d+e-f)$	*21*	$(2f-g+h)(2f+g-h)$
22	$(1-x+y)(1+x-y)$	*23*	$(2-m+n)(2+m-n)$	*24*	$(3-a-b)(3+a+b)$

25 $4ab$ *26* $-4pq$ *27* $(x+3y)(7x-3y)$ *28* $(3a-2b)(7a+2b)$

Page 8 Exercise 3

1 32 *2* 96 *3* 39 *4* 2600 *5* 34 *6* 46 *7* 9800 *8* 998 000

9 7 *10* 9 *11* 12 *12* 11 *13a* 50 *b* 400 *c* 36 *d* 0·352

14 $\pi(R-r)(R+r)$ *a* 628 cm^2 *b* 220 cm^2

15 $A=\pi(R-10r)(R+10r)$; 3140 cm^2 *16* $\frac{1}{2}m(v-u)(v+u)$; 24 000

17a $(\cos x-\sin x)(\cos x+\sin x)$ *b* $(1-\tan x)(1+\tan x)$

c $(2\sin x-1)(2\sin x+1)$ *d* $(3-2\cos x)(3+2\cos x)$ *18b* 2880

Page 11 Exercise 4

1 $x^2+7x+12$ *2* x^2+8x+7 *3* $x^2-7x+10$ *4* $x^2-12x+36$

5 3, 2 *6* 4, 5 *7* 1, 7 *8* 3, 3 *9* 1, 10

10 $-2, -5$ *11* $-3, -5$ *12* $-5, -6$ *13* $-1, -2$ *14* $-1, -8$

15 $(x+4)(x+3)$ *16* $(x+1)(x+5)$ *17* $(x-4)(x-2)$ *18* $(x-1)(x-2)$

19 $(x+3)(x+5)$ *20* $(x-2)(x-10)$ *21* $(x+1)(x+2)$ *22* $(a-1)(a-2)$

23 $(x+4)(x+2)$ *24* $(p-2)(p-4)$ *25* $(x+3)(x+6)$ *26* $(q-3)(q-6)$

27 $(k-2)(k-6)$ *28* $(a+2)(a+6)$ *29* $(x+1)(x+10)$ *30* $(z+3)(z+7)$

31 $(c-1)(c-8)$ *32* $(n+4)(n+8)$ *33* $(x+3)(x+8)$ *34* $(v-2)(v-12)$

35 $(b+5)(b+5)$ *36* $(m+4)(m+7)$ *37* $(x-5)(x-9)$ *38* $(d+3)(d+15)$

39 $(x-4)(x-15)$ *40* $(y+4)(y+20)$ *41* $(3+x)(2+x)$ *42* $(5-x)(3-x)$

43 $(10+x)(5+x)$ *44* $(2-x)(18-x)$ *45* $(4+x)(4+x)$ *46* $(5-a)(5-a)$

47 $(1-b)(1-b)$ *48* $(3+x)(8+x)$

Page 12 Exercise 5

1 $x^2+3x-18$ *2* x^2+7x-8 *3* $x^2-3x-10$ *4* $x^2-9x-10$

5 $-4, 5$ *6* $-1, 7$ *7* $-4, 4$ *8* $-3, 3$ *9* $5, -1$

10 $-2, 5$ *11* $-6, 3$ *12* $-8, 3$ *13* $-18, 2$ *14* $-2, 1$

15 $(x+6)(x-2)$ *16* $(x+1)(x-9)$ *17* $(x-2)(x+1)$ *18* $(x-12)(x+3)$

19 $(a+5)(a-2)$ *20* $(b-5)(b+2)$ *21* $(c+7)(c-1)$ *22* $(d-7)(d+1)$

23 $(x+9)(x-4)$ *24* $(x-9)(x+4)$ *25* $(x+6)(x-3)$ *26* $(x-6)(x+3)$

27 $(m-6)(m+4)$ *28* $(n-8)(n+3)$ *29* $(p-10)(p+3)$ *30* $(q-9)(q+5)$

31 $(a+2)(a-1)$ *32* $(b+3)(b-2)$ *33* $(c-4)(c+3)$ *34* $(d-5)(d+4)$

35 $(p+5)(p-3)$ *36* $(q+7)(q-3)$ *37* $(r+7)(r-1)$ *38* $(s+6)(s-2)$

39 $(t+4)(t-2)$ *40* $(u-4)(u+2)$ *41* $(v+8)(v-1)$ *42* $(w-8)(w+1)$

43 $(x+12)(x-2)$ *44* $(x-7)(x+6)$ *45* $(a+8)(a-6)$ *46* $(a-12)(a+6)$

47 $(y+8)(y-9)$ *48* $(y-24)(y+3)$ *49* $(z+18)(z-4)$ *50* $(z+72)(z-1)$

Page 14 Exercise 6

1 $(2x+3)(x+1)$ *2* $(2x+1)(x+3)$ *3* $(3x+1)(x+2)$ *4* $(2a+1)(3a+2)$

5 $(5x+1)(2x+3)$ *6* $(3x+5)(x+3)$ *7* $(3c+1)(3c+1)$ *8* $(2x+3)(5x+2)$

9 $(6y-1)(2y-1)$ *10* $(2n-1)(n-3)$ *11* $(2z-1)(z-2)$ *12* $(2x-3)(x-1)$

13 $(4b-5)(2b-1)$ 14 $(3a-2)(2a-3)$ 15 $(3c-4)(3c-4)$ 16 $(2p-1)(p-9)$

17 $(4y-5)(3y-2)$ 18 $(10m-3)(m-4)$ 19 $(3z+4)(z-2)$ 20 $(4p-1)(p+2)$

21 $(3n+1)(n-2)$ 22 $(3k-2)(k+1)$ 23 $(3n+2)(2n-3)$ 24 $(5y-1)(y+1)$

25 $(6x-1)(x+3)$ 26 $(2x-1)(x+1)$ 27 $(3x+1)(x-1)$ 28 $(5x+2)(x-5)$

29 $(4x+3)(2x-1)$ 30 $(4c-1)(2c+3)$ 31 $(3z-1)(2z-3)$ 32 $(4a-5)(3a+1)$

33 $(2x+3)(2x+3)$ 34 $(6y-1)(4y+1)$ 35 $(1+6x)(1-3x)$ 36 $(5+p)(3-2p)$

Page 14   Exercise 6B

1   $(a+2)(a+6)$ 2 $(a+2b)(a+6b)$ 3 $(p+2q)(p+6q)$

4   $(c-4)(c-6)$ 5 $(x-4y)(x-6y)$ 6 $(c-4d)(c-6d)$

7   $(a+b)(a+2b)$ 8 $(p-2q)(p+q)$ 9 $(r+3s)(r-2s)$

10   $(u-v)(u-v)$ 11 $(a-7b)(a+2b)$ 12 $(c+10d)(c-3d)$

13   $(2x+y)(x+5y)$ 14 $(2y-z)(y+z)$ 15 $(4a+b)(a-2b)$

16   $(2a+b)(a-3b)$ 17 $(2a-3b)(a-b)$ 18 $(2a-b)(a+3b)$

19   $(3x-2y)(3x+4y)$ 20 $(3r+2s)(2r-3s)$ 21 $(4x-y)(x-4y)$

22   $(4a-b)(3a+4b)$ 23 $(9x+10y)(x-y)$ 24 $(8a+15b)(a-b)$

25   $(2\cos\theta-1)(\cos\theta+2)$ 26 $(\sin\theta-1)(3\sin\theta-1)$

27   $(5\tan\theta-3)(\tan\theta+1)$ 28 $(3\sin\theta-2)(3\sin\theta-2)$

29   $(2\tan\theta+5)(3\tan\theta-4)$ 30 $(5\cos\theta-3)(2\cos\theta+1)$

Page 15   Exercise 7

1   $5(x+3y)$ 2 $(x-5)(x+5)$ 3 $(a+3)(a+3)$

4   $x(x-1)$ 5 $(y+2)(y-3)$ 6 $(1-p)(1+p)$

7   $(a+2)(a+2)$ 8 $a(x-p+d)$ 9 $3(x-2)(x+2)$

10   $(p-q)(p+q)$ 11 $p(p-2)$ 12 $(p-1)(p-1)$

13   $(a-1)(a+1)$ 14 $a(a-1)$ 15 $(a+1)(a-2)$

16   $2(x-3)(x+3)$ 17 $2x(x-4)$ 18 $x^2(x-1)$

19   $(2k+5)(k-1)$ 20 $(k-3)(k-3)$ 21 $(3k+1)(3k+1)$

22   $(4-x)(4+x)$ 23 $2(3-y)(3+y)$ 24 $6z(2-z)$

25   $(3-2a)(3+2a)$ 26 $(2p+1)(p-1)$ 27 $(2x+3)(3x+2)$

28   $7(2a^2+3b^2)$ 29 $14(a-2b)(a+2b)$ 30 $(4x-1)(4x-1)$

31   $4ab(a-2b)$ 32 $(1-x)(1-x)$ 33 $x(a+b-2)$

34 $2x(x-4)(x+4)$ 35 $(3y-2)(2y+3)$ 36 $2(x+1)(x+1)$

37 $3(a+2)(a-1)$ 38 $2(2b-1)(b+4)$ 39 $5x(1-2x)(1+2x)$

40 $6x(1+4x^2)$ 41 $k(m-n)(m+n)$ 42 $(2c+3)(c-5)$

43 $a^2(1+a+a^2)$ 44 $(4x-3)(x-2)$

Page 16 Exercise 7B

1 $x^4(1-x)(1+x)$

2 $(x-1)(x+1)(x^2+1)$

3 $(a-b-x)(a-b+x)$

4 $(a-5b)(a-5b)$

5 $(m-n)(m+n)(m^2+n^2)$

6 $4a^2b^2(b-3a)$

7 $(1-p+q)(1+p-q)$

8 $(3a+4b)(4a-3b)$

9 $(4-k)(4-k)$

10 $x^2y^2(y-x)(y+x)(y^2+x^2)$

11 $2x(x-1)(x+2)$

12 $3(1+3x)(1-4x)$

13 $(x-y-z)(x+y+z)$

14 $x(x-6)$

15 $(2a+3b)(3a-5b)$

16 $2\tan A(\tan A-2)$

17 $(2\cos A-\sin A)(2\cos A+\sin A)$

18 $(2\sin A+3)(\sin A+1)$

19 $2(1-x+y)(1+x-y)$

20 $(x-y)(a-b)$

21 $(p-q-r+s)(p-q+r-s)$

22 $x(x-3)(x+3)(x^2+9)$

23 $(x-y)(x-y-3)$

24 $(p-q)(a-1)$

25 $(x-1)(x+1)$

26 $(x+1)(x-1)(x+3)$

27 $2(a-b)(1-3a+3b)(1+3a-3b)$

28 $(6a-5)(2a+3)$

29 $(1-x)(1+x)(1+x^2)(1+x^4)$

30 $3a^3b^3(2b-3a^2+ab^2)$

31 $r(r-2)$

32 $4r^3$

33 $n(n+1)(n^2+n-1)$

Page 17 Exercise 8

1 $\frac{3}{5}$ 2 $\frac{7}{15}$ 3 $\frac{1}{2}$ 4 $\frac{1}{3}$ 5 $\frac{2}{3}$

6 $2x+3$ 7 $3a+2$ 8 $7a+5b$ 9 $2p-4q$ 10 $4x-8y$

11 $3x-2y$ 12 $b+c$ 13 $a-3b$ 14 $4a+b$ 15 $3a+2a^2$

16 $\dfrac{a+b}{2c}$ 17 $\dfrac{y-z}{2}$ 18 $\dfrac{1}{2a+3b}$ 19 $\dfrac{1}{2a+3b}$ 20 $\dfrac{3}{2y-z}$

21 $\dfrac{q}{3r+q}$ 22 $\dfrac{1}{a-b}$ 23 $\frac{1}{7}$ 24 $4a$ 25 $2a-3$

26    $a+2$ 27 $x+y$ 28 $\dfrac{1}{x+1}$ 29 $\dfrac{3}{p-1}$ 30 $\dfrac{a}{x+3}$

31   $\frac{4}{3}$ 32 $\dfrac{x-1}{x+1}$ 33 $\dfrac{3a+1}{a+2}$ 34 $\dfrac{x-3y}{x+2y}$

35   $\dfrac{x+1}{x+3}$ 36 $\dfrac{a-5}{a+7}$ 37 $\dfrac{2c-3}{c+4}$

38a   T   b   F   c   T   d   T   e   F   f   T

Page 19   Exercise 8b

1   -1 2 -2 3 1 4 $-a$ 5 $-\dfrac{a}{d}$

6   $\dfrac{a}{c}$ 7 $-(x+2)$ 8 $x-2$ 9 $-(x+y)$

10   $1-x$ 11 $2x+1$ 12 $-\frac{1}{2}(x+3)$ 13 $\dfrac{a-2}{a}$

14   $\dfrac{1+x}{2+x}$ 15 $-(x^2+1)$ 16 $1-\cos A$ 17 $\frac{1}{2}(2-\sin A)$

18   $-\dfrac{3}{1+\sin A}$ 19 $\dfrac{1+\cos x}{1-\cos x}$ 20 $\dfrac{\cos x-1}{\cos x+4}$ 21 $\frac{1}{7}$ 22 $\frac{9}{8}$

23   $\dfrac{3a+1}{3a-1}$ 24 $\dfrac{ab+1}{ab-1}$ 25 $\dfrac{x-1}{x+1}$ 26 $\dfrac{1}{ab}$

27   $\dfrac{c+1}{c}$ 28 $\dfrac{y}{y-x}$ 29 $1-a$ 30 $\dfrac{1+2a}{a}$

31   $-\dfrac{1}{x(x+h)}$ 32 $\dfrac{2sc}{c^2+s^2}$ 33 $\dfrac{x-1}{x+1}$ 34 $\dfrac{x+2}{x-1}$

35   $\dfrac{a^3}{a-b}$ 36a $2\left(1-\dfrac{x}{2}\right)$ b $4\left(1+\dfrac{x}{4}\right)$ c $3\left(x-\dfrac{2}{3}\right)$

d $4\left(a-\dfrac{x}{2}\right)$ e $a\left(x+\dfrac{b}{a}\right)$ f $2x\left(1-\dfrac{3}{2}x\right)$

Page 22   Exercise 9

1 a $\dfrac{4}{5}$ b $\dfrac{4a}{5}$ c $\dfrac{4}{5a}$ 2 a $\dfrac{3}{4}$ b $\dfrac{3a}{4}$ c $\dfrac{3}{4a}$ 3 a $\dfrac{2}{5}$ b $\dfrac{2a}{5}$ c $\dfrac{2}{5a}$

4 a $\dfrac{1}{6}$ *b* $\dfrac{a}{6}$ *c* $\dfrac{1}{6a}$ *5 a* $\dfrac{13}{12}$ *b* $\dfrac{13a}{12}$ *c* $\dfrac{13}{12a}$ *6 a* $\dfrac{1}{12}$ *b* $\dfrac{a}{12}$ *c* $\dfrac{1}{12a}$

7 $\dfrac{x}{6}$ *8* $\dfrac{1}{6x}$ *9* $\dfrac{3}{a}$ *10* $\dfrac{a+2}{b}$ *11* $\dfrac{2}{a}$ *12* $\dfrac{1}{a}$.

13 $\dfrac{3}{2a}$ *14* $\dfrac{7}{2a}$ *15* $\dfrac{5}{9a}$ *16* $\dfrac{a}{4c}$ *17* $\dfrac{x+y}{xy}$ *18* $\dfrac{u-v}{uv}$

19 $\dfrac{2d+3c}{cd}$ *20* $\dfrac{pq-12}{3q}$ *21* $\dfrac{ay+x}{xy}$ *22* $\dfrac{ay+bx}{xy}$ *23* $\dfrac{ad-bc}{bd}$ *24* $\dfrac{a^2+b^2}{ab}$

25 $\dfrac{ab-a^2}{b^2}$ *26* $\dfrac{2a}{3}$ *27* $\dfrac{9}{a}$

28 $\dfrac{6}{a}$ *29* $\dfrac{23x}{12}$ *30* $\dfrac{a^2+a+1}{a^2}$ *31* $\dfrac{1-c-c^2}{c^2}$ *32* $\dfrac{bc+ca+ab}{abc}$

33 $\dfrac{x^2+1}{x}$ *34* $\dfrac{x-1}{x}$ *35* $\dfrac{ab+c}{b}$ *36* $\dfrac{3x+2}{4}$ *37* $\dfrac{x-1}{12}$

38 $\dfrac{3x-1}{10}$ *39* $\dfrac{5(x-2)}{12}$ *40* $\dfrac{3}{2}$ *41* $\dfrac{c}{15}$

Page 23   Exercise 9B

1 $\dfrac{2x}{(x+1)(x-1)}$ *2* $\dfrac{5a+7}{(a+2)(a+1)}$ *3* $\dfrac{9x+y}{(x+y)(x-y)}$ *4* $\dfrac{10}{x(x+1)}$

5 $\dfrac{5}{a(a-1)}$ *6* $-\dfrac{(a+5b)}{(a+b)(a-b)}$ *7* $\dfrac{x^2-2xy-y^2}{(x+y)(x-y)}$ *8* $\dfrac{a^2+b^2}{(a-b)(a+b)}$

9 $\dfrac{1}{x-1}$ *10* $\dfrac{a}{(1-a)(a+1)}$ *11* $\dfrac{x+4}{(x+1)(x+3)}$ *12* $\dfrac{a+2}{(a+1)^2}$

13 $\dfrac{y}{(y+3)(y-2)}$ *14* $\dfrac{2a}{(a+1)(a-1)^2}$ *15* $\dfrac{1}{(x-3)(x+3)}$ *16* $\dfrac{.2y+1}{(y-2)(y+2)}$

17 $\dfrac{x+5}{(x-1)(x+1)(x+2)}$ *18* $-\dfrac{2x+5}{(x+2)(x+3)}$ *19* $\dfrac{2(x^2+1)}{(x+1)(x-1)}$ *20* $\dfrac{1}{x^2(x+1)}$

21 $\dfrac{2c-1}{c^2(c-1)}$ *22* $\dfrac{5}{(3x-4)(3x+4)(2x+1)}$ *23* $\dfrac{\sin A+\cos A}{\sin A\cos A}$

24 $\dfrac{1}{\sin A\cos A}$ *25* $\dfrac{2}{(1-\sin A)(1+\sin A)}$ *26* $\dfrac{\sin x-\cos x-1}{(1+\cos x)(1-\sin x)}$

Algebra—Answers to Chapter 2

Page 29   Exercise 1

1 a 48, 5   *b* 18, 1·2   *c* 6, 9, 12

2 (2, 8), (5, 20), (8, 32), (10, 40), (1·75, 7); $k = 4$, $y = 4x$

3 9, 10·5, −6, 13·5, 14, −8   *4 a* $P \propto x$, $P = kx$ *b* $V \propto T$, $V = kT$

  c $c \propto r$, $c = kr$ *d* $s \propto t$, $s = kt$ *e* $P \propto d$, $P = kd$

5 $k = \frac{5}{2}$, 25   *6* $k = 2$, 22   *7* $k = \frac{1}{2}$, 4·5   *8* $y = 4x$ *a* 36   *b* 13

9 $k = \frac{3}{2}$ *a* 36   *b* 22; trebled also   *10* 12   *11* $T = \frac{16}{3}e$; 12·8

12 $P = \frac{6}{25}W$ *a* 28·8   *b* 300   *13* $m = 2·8t$; 7

14b $V = 20I$ *c* (*1*) 9   (*2*) 0·32

Page 32   Exercise 2

1 42·5, 45, 47·5, 50, 52·5, 55; 25, 29·6   *2* 26·7, 29·3, 32, 34·7, 37·3; 7·8, 8·7

3 8·05, 16·1, 24·2, 32·2, 40·3, 48·3, 161

4 34, 54, 55, 62·5, 65, 67·5, 71, 79, 95   *5* 125; 0·7

6 a 12·2   *b* 9·5   *7 a* 7·77   *b* 6·19   *8* {1·04, 2·61, 3·60}

9 2·20, 4·40, 6·60, 8·80   *10* 6·3; 3·75

Page 36   Exercise 3

1 a $p = kq^2$ *b* $r = k\sqrt{A}$ *c* $W = kd^3$ *d* $D = k\tan x$

2 a p varies as the square of q, etc.   *3* $k = 3$ *a* 75   *b* 0·4

4 a 3   *b* 100   *5 a* 500   *b* 3   *6* 40

9 7200   *10* 16 cm   *11* 2 m   *12* 19·5 m/s

13 180 m   *14* 3 s   *15* 1·4 s   *16* 1·84 years

17a 9 : 4   *b* 10

18a y is multiplied by 4   *b* y is reduced to $\frac{1}{9}$ of its value

19a y is multiplied by 8   *b* y is reduced to $\frac{1}{27}$ of its value

20 $y = 2x^2$; (1, 2), (2, 8), (3, 18), (5, 50)   *21b* $V = 14·2\sqrt{h}$ *c* 63·5

1 y varies inversely as x or x varies inversely as y, the constant of variation being 4 in both cases   *a* 2   *b* 40

2 $y = \dfrac{12}{x}$; 2   *3* $y = \dfrac{84}{x}$; 3   *4* 2·5   *5* 12·5

6 a $y = \dfrac{24}{x}$ *b* 0·4   *c* 1·6   *7 a* 400   *b* 1·5   *8 a* $H = \dfrac{6400}{R}$ *b* 13·3

9 a $f = \dfrac{300\,000}{\lambda}$ *b* 1500 m   *c* 647 kHz   *10c* $k = 6$ *d* 1·2

1 a $p = \dfrac{k}{v}$ or $pv = k$ *b* $y = \dfrac{k}{\sqrt{x}}$ or $y\sqrt{x} = k$ *c* $y = \dfrac{k}{x^3}$ or $x^3 y = k$

2 a p varies inversely as v *b* y varies inversely as the square root of x
   c y varies inversely as the cube of x

3 a 0·36   *b* ± 3 *4* 3   *5* 27   *6 a* 2   *b* $\frac{1}{9}$

7   0·08   *8 a* $I = \dfrac{25}{d^2}$ *b* 4   *c* $\sqrt{10}$

9   12·5 cm   *10a* 0·25   *b* 0·16

11 $W = \dfrac{k}{x^2}$ and $W_o = \dfrac{k}{R^2} \Rightarrow W = \dfrac{R^2 W_o}{x^2}$; 256 newtons

12a y is quartered   *b* y is quadrupled (multiplied by 4)

1 a $A = krh$ *b* $p = \dfrac{kt}{v}$ *c* $Q = kxy^2$ *d* $F = \dfrac{km_1 m_2}{d^2}$

2 a $x = 4yz$ *b* 80   *3 a* $y = 2xz$ *b* 40   *c* 1

4   7·5   *5 a* $F = \dfrac{6y}{z}$ *b* 15   *6* $6\frac{1}{4}$ or 6·25   *7* 20

8 a 7·5   *b* 24   *9* $W = \frac{1}{12}Ld^2$ *a* 1080   *b* 180   *c* 6   *10* $1\frac{1}{3}$

1 a $y = \dfrac{7xz}{2}$ *b* 63   *2* 7·2   *3* $y = \dfrac{10x^2}{\sqrt{z}}$; 90   *4* 1·5

5 a $V = \dfrac{920T}{p}$ *b* 299   *6·a* $p = 250dr^2$ *b* 2250Pa   *c* 1·5 m

288 ANSWERS

7 *a* 172·8 kN   *b* 30 m/s   *8* 3   *9* 200

10a P is doubled *b* 50% increase

11 60% increase   *12a* 5376 *b* 70% decrease

Page 49 Exercise 7

1 (i) $y = 2x + 1$ (ii) $y = 6 - \frac{1}{2}x$

2 a 1·5, 4 *b* −3, 13 *c* 0·4, 1·3 *d* −0·8, 20·5

3 a $y = 0\cdot6x + 1\cdot5$ *b* $y = 50 - 1\cdot4x$ *4* $P = 0\cdot06L + 0\cdot40$

5 a $R = 0\cdot04t + 10\cdot9$ *b* $R = 11\cdot6$ *6 b* $v = 30 - 0\cdot8t$ *c* (*1*) 23·6 (*2*) 37·5

7 $V = 24x + 40$ *8* $Q = 500 - 2\cdot5t$; 3 min 20 s

Algebra—Answers to Chapter 3

Page 53 Exercise 1

1 a 0 *b* 0   *2 a* 1 *b* 5 *c* −4 *d* −3 *e* $\frac{1}{2}$ *f* $-\frac{2}{3}$

3 a 0   *b* 0   *c* yes

4 a 1, 2 *b* 5, −5 *c* −1, $\frac{1}{2}$ *d* $\frac{3}{2}, \frac{2}{3}$ *e* 0, −7

 f 0, 9 *g* 0, $\frac{1}{2}$ *h* 0   *i* 1, −5

5 $\{-2, 4\}$ *6* $\{3, 8\}$ *7* $\{3, -\frac{1}{2}\}$ *8* $\{-10, 10\}$ *9* $\{\frac{1}{2}, -\frac{3}{2}\}$

10 $\{-\frac{1}{3}, \frac{1}{3}\}$ *11* $\{0, -4\}$ *12* $\{0, 10\}$ *13* $\{0, \frac{2}{5}\}$ *14* $\{0, \frac{1}{3}\}$

15 $\{0, 7\}$ *16* $\{0, -\frac{3}{2}\}$ *17* $\{\frac{1}{2}, \frac{2}{3}\}$ *18* $\{2, -\frac{1}{5}\}$ *19* $\{6, -3\}$

20 $\{\frac{5}{2}, 2\}$ *21* $\{-7, 7\}$ *22* $\{\frac{5}{2}\}$ *23* $\{1, 2, 3\}$ *24* $\{-2, \frac{1}{2}, 3\}$

Page 55 Exercise 2

1 $\{0, -2\}$ *2* $\{0, 5\}$ *3* $\{0, -8\}$ *4* $\{0, \frac{1}{2}\}$

5 $\{0, \frac{2}{5}\}$ *6* $\{0, \frac{3}{4}\}$ *7* $\{0, 2\}$ *8* $\{0, -3\}$

9 $\{0, 3\}$ *10* $\{-3, 3\}$ *11* $\{-\frac{1}{2}, \frac{1}{2}\}$ *12* $\{-\frac{2}{3}, \frac{2}{3}\}$

13 $\{-\frac{7}{2}, \frac{7}{2}\}$ *14* $\{-1, 1\}$ *15* $\{-\frac{3}{2}, \frac{3}{2}\}$ *16* $\{-4, 4\}$

17 $\{-2, 2\}$ *18* $\{-12, 12\}$ *19* $\{0\}$ *20* $\{0\}$

21 $\{2\}$ *22* $\{1, 2\}$ *23* $\{3, 4\}$ *24* $\{1, 5\}$

25 $\{-2, 3\}$ *26* $\{-5, 3\}$ *27* $\{-1, -4\}$ *28* $\{-3, -7\}$

29 $\{-3, -1\}$ *30* $\{3\}$ *31* $\{-10\}$ *32* $\{\frac{1}{2}, 2\}$

33 $\{-3, \frac{1}{2}\}$ *34* $\{1, 1\frac{1}{2}\}$ *35* $\{\frac{1}{3}, 3\}$ *36* $\{-\frac{1}{3}, 3\}$

37 $\{-\frac{4}{5}, 1\}$ *38* $\{\frac{7}{2}, -5\}$ *39* $\{\frac{3}{4}, 4\}$ *40* $\{-5, 3\}$

41 $\{-\frac{4}{3}\}$ *42* $\{-\frac{2}{3}, \frac{3}{2}\}$ *43* $\{-5, -\frac{1}{2}\}$ *44* $\{\frac{2}{3}\}$

45 $\{-\frac{3}{4}, \frac{4}{3}\}$

Page 57 Exercise 2B

1 $\{-2, -\frac{1}{2}\}$ *2* $\{-\frac{3}{2}, -\frac{1}{2}\}$ *3* $\{-2, \frac{5}{2}\}$ *4* $\{-\frac{2}{3}, \frac{7}{2}\}$

5 $\{-4, \frac{3}{2}\}$ *6* $\{\frac{1}{2}, \frac{5}{3}\}$ *7* $\{\frac{1}{3}, -\frac{7}{2}\}$ *8* $\{-\frac{7}{2}, \frac{3}{2}\}$

9 $\{-\frac{2}{3}\}$ *10* $\{-2\}$ *11* $\{3\}$ *12* $\{-7, 4\}$

13 $\{-\frac{7}{2}, 1\}$ *14* $\{-2, 12\}$ *15* $\{-\frac{1}{3}, \frac{3}{2}\}$ *16* $\{-8, 4\}$

17 $\{-3, 8\}$ *18* $\{-\frac{5}{2}, \frac{3}{2}\}$ *19* $\{-2, \frac{7}{2}\}$ *20* $\{1, 2\}$

21 $\{-\frac{1}{3}, 1\}$ *22* $\{0, 1\}$ *23* $\{-1, 1\}$ *24* $\{-1, 6\}$

25 $\{-1, -3\}$ *26* $\{-\frac{5}{2}, 3\}$ *27* $\{-5, -3\}$ *28* $\{\frac{1}{2}, 4\}$

29 $\{\frac{8}{3}, \frac{9}{2}\}$ *30* $\{-3, -1\}$ *31* $\{-6, 3\}$ *32* $\{-4, \frac{5}{2}\}$

33 $\{-1, 6\}$ *34* $\{1, 2\}$ *35* $\{2\}$ *36* $\{-1, 2\}$

37 $\{-1, 9\}$ *38* $\{-\frac{3}{2}, 4\}$ *39* $\{-2, 2\}$ *40* $\{-6, 5\}$

41 $\{-3, \frac{9}{2}\}$ *42* $\{0, 11\}$ *43* $\{\frac{1}{3}, 2\}$ *44* $\{-3, 2\}$

45 $\{0, 3\}$ *46* $\{0, 2\}$ *47* $\{-8, 7\}$ *48* $\{-2, 5\}$

49 $\{-3, 5\}$ *50* $\{-\frac{4}{3}, -\frac{6}{7}\}$

Page 59 Exercise 3

1 6, 13   *2* 12, 14   *3 a* $25-n$ *b* 8, 17   *4* 15 m by 12 m

5 3   *6* 20 cm by 15 cm   *7 a* $(80-x)$ cm   *b* 50 cm by 30 cm

8 after 1·8 seconds and again after 3 seconds; going up and going down.

9 16 m by 5 m   *10* 15 cm   *11* 20

12 13   *13a* 9   *b* 4   *c* (6, 0)

14a A$(-2, 0)$, B$(6, 0)$ *b* P$(3, -15)$ *15* 3; 6 cm²

Page 61 Exercise 3B

1 1   *2* $\frac{2}{5}$ or $\frac{5}{2}$ *3* $\frac{1}{2}$ or 5

4 a $(-4, 0)$, $(-3, 0)$ *b* (0, 2), (0, 6)   *5* 3 or 9

6 a $(x-t)$ cm   *b* $A = \pi x^2 - \pi(x-t)^2$, etc.   *c* 3   (7 impossible)

7 a $(12-2x)$ cm, $(9-2x)$ cm   *b* $(12-2x)(9-2x) = 54$; width of strip $= 1·5$ cm

8 a $(x-2)$ cm, $(x-2)$ cm, $(x-1)$ cm   *b* $V = x^3 - (x-2)^2(x-1)$, etc.   *c* 10

9 a $\dfrac{20}{15-x}$ hours, $\dfrac{20}{15+x}$ hours   *b* $\dfrac{20}{15-x} + \dfrac{20}{15+x} = 3$; 5 km/h

10a $(x-1)$ cm, $(x+8)$ cm   *b* 20, 21, 29 cm

11a $(2\cdot5 - d)$ cm   *c* $\frac{1}{2}$ *12b* 8

13a $8(x^2-1) = 2x-5$, etc.   *b* A$(-\frac{1}{2}, -6)$, B$(\frac{3}{4}, -\frac{7}{2})$

14 $x = 5$ or $x = -1$; P$(-1, 7)$, Q$(5, -5)$

Page 65   Exercise 4B

1 $\{-6, 6\}$ *2* $\{-10, 10\}$ *3* ϕ *4* $\{-\sqrt{3}, \sqrt{3}\}$

5 $\{-\frac{1}{2}, \frac{1}{2}\}$ *6* $\{-\frac{3}{4}, \frac{3}{4}\}$ *7* $\{-\frac{1}{2}, \frac{1}{2}\}$ *8* $\{-\frac{5}{2}, \frac{5}{2}\}$

9 $\{-1, 3\}$ *10* $\{-3, 1\}$ *11* $\{-7, 13\}$ *12* $\{-2, 6\}$

13 $\{-8, -2\}$ *14* $\{-14, -4\}$ *15* $\{-2, 6\}$ *16* $\{-4\frac{1}{2}, 5\frac{1}{2}\}$

17 $\{-1\frac{1}{2}, \frac{1}{2}\}$ *18* $\{-3\frac{1}{4}, 2\frac{3}{4}\}$ *19* $\{-5\frac{1}{4}, 6\frac{3}{4}\}$ *20* $\{-4, 5\}$

21 $\{-\frac{5}{3}, 3\}$ *22* $\{0, 5\}$ *23* $\{\frac{4}{3}, 2\}$ *24* $\pm 3y$

25 $\pm 2\sqrt{y}$ *26* $\pm\sqrt{(2y)}$ *27* $\pm\frac{1}{2}y$ *28* $\pm\sqrt{(a^2 - y^2)}$

29 $\pm\sqrt{(a^2 - b^2)}$ *30* $\pm\frac{1}{2}\sqrt{(p^2 + q^2)}$ *31* $\pm\dfrac{b}{\sqrt{(a^2 - c^2)}}$ *32* $\pm\dfrac{1}{\sqrt{(a^2 - 1)}}$

33 $r = \sqrt{\dfrac{A}{\pi}}$ *34* $r = \sqrt{\dfrac{V}{\pi h}}$ *35* $v = \sqrt{\dfrac{2T}{m}}$ *36* $r = \sqrt{\dfrac{3V}{\pi h}}$

37 $v = \sqrt{\dfrac{Fr}{m}}$ *38* $D = \sqrt{\dfrac{kL}{R}}$ *39a* $R = \sqrt{\dfrac{(A + \pi r^2)}{\pi}}$ *b* $r = \sqrt{\dfrac{(\pi R^2 - A)}{\pi}}$

40a $R = \sqrt{\dfrac{(V + \pi h r^2)}{\pi h}}$ *b* $r = \sqrt{\dfrac{(\pi h R^2 - V)}{\pi h}}$ *41* $e = \dfrac{\sqrt{(a^2 - b^2)}}{a}$

42 $k = \dfrac{\sqrt{(T^2 - c^2)}}{c}$

Page 67   Exercise 5B

1 4   *2* 16   *3* 25   *4* 9   *5* 1   *6* 36

7 $\frac{1}{4}$ *8* $\frac{9}{4}$ *9* $\frac{1}{9}$ *10* $\frac{49}{4}$ *11* $\frac{81}{4}$ *12* $\frac{1}{16}$

13 4; 2   *14* 9; 3   *15* 36; 6   *16* 49; 7   *17* $\frac{1}{4}; \frac{1}{2}$ *18* $\frac{1}{4}; \frac{1}{2}$

19 0·09; 0·3   *20* $\frac{25}{4}; \frac{5}{2}$ *21* $\{-4, 2\}$

22 $\{-1, 3\}$ *23* $\{-3, 7\}$ *24* $\{1, 3\}$

25 $\{-9, 3\}$ 26 $\{-2, 8\}$ 27 $\{-12, 2\}$

28 $\{6, 8\}$ 29 $\{-5\}$ 30 $\{-2, 1\}$

31 $\{\ \}$ or ø   32 $\{-2, -3\}$ 33 $\{-8, 5\}$

34 $\{-2, \frac{3}{2}\}$ 35 $\{-\frac{7}{2}, 2\}$ 36 $\{-3\cdot7, -0\cdot3\}$

37 $\{5\cdot4, 0\cdot6\}$ 38 $\{-5\cdot2, -0\cdot8\}$ 39 $\{-0\cdot4, 2\cdot4\}$

40 $\{-8\cdot1, 0\cdot1\}$ 41 $\{-2\cdot6, -0\cdot4\}$ 42 $\{-0\cdot6, 3\cdot6\}$

43 $\{-1\cdot4, 0\cdot4\}$ 44 $\{-2\cdot3, 0\cdot3\}$ 45 $\{-0\cdot6, 6\cdot6\}$

46 $\{-1\cdot1, 7\cdot1\}$ 47 $\{-1\cdot1, 2\cdot6\}$ 48 $\{-0\cdot4, 1\cdot9\}$

49 $\{-1\cdot1, 4\cdot1\}$ 50 $\{29\cdot8, 0\cdot2\}$

Page 69   Exercise 6

1 $\{-3, -2\}$ 2 $\{1, 2\}$ 3 $\{-3, -\frac{1}{2}\}$

4 $\{-3, 1\frac{1}{2}\}$ 5 $\{1\frac{1}{2}, 4\}$ 6 $\{-1\frac{1}{3}, 2\}$

7 $\{-4, \frac{2}{3}\}$ 8 $\{-\frac{1}{2}, 2\frac{1}{2}\}$ 9 $\{2\frac{1}{2}\}$

10 $\{-5, \frac{2}{3}\}$ 11 $\{-7, \frac{1}{3}\}$ 12 $\{-1\frac{1}{4}, 5\}$

13 $\{-3\cdot7, -0\cdot3\}$ 14 $\{-5\cdot2, -0\cdot8\}$ 15 $\{-6\cdot2, -0\cdot8\}$

16 $\{-2\cdot4, 0\cdot4\}$ 17 $\{0\cdot6, 5\cdot4\}$ 18 $\{-1\cdot3, 5\cdot3\}$

19 $\{-9\cdot1, 1\cdot1\}$ 20 $\{-0\cdot4, 12\cdot4\}$ 21 $\{-19\cdot2, -0\cdot8\}$

22 $\{-1\cdot7, 1\cdot2\}$ 23 $\{-0\cdot8, 2\cdot4\}$ 24 $\{-5\cdot1, -0\cdot9\}$

25 $\{0\cdot3, 2\cdot7\}$ 26 $\{1\cdot4, 2\cdot6\}$ 27 $\{-1\cdot2, 0\cdot6\}$

28 $\{-0\cdot5, 2\cdot9\}$ 29 $\{-0\cdot1, 0\cdot8\}$ 30 $\{-0\cdot6, 0\cdot4\}$

31 $\{0\cdot6, 5\cdot4\}$ 32 $\{-2\cdot8, 0\cdot4\}$ 33 $\{-1\cdot7, 0\cdot5\}$

34 $\{-1\cdot3, 11\cdot3\}$ 35 $\{-1\cdot0, 4\cdot0\}$ 36 $\{-6\cdot2, -1\cdot3\}$

37 $\{-0\cdot7, 2\cdot0\}$ 38 $\{1\cdot1, 0\cdot1\}$ 39 $\{4\cdot6, -0\cdot6\}$

40 $\{-0\cdot6, 2\cdot1\}$ 41 $\{0\cdot4, 2\cdot6\}$ 42 $\{0\cdot6, 5\cdot4\}$

43a $\dfrac{-5\pm\sqrt{(25-4a)}}{2}$ b $\dfrac{a\pm\sqrt{(a^2-72)}}{4}$

c $\dfrac{-1\pm\sqrt{(1-a^2)}}{a}$ d $-2\pm\sqrt{(a-1)}$

44 $7x^2-14x-9=0$; $\{-0\cdot5, 2\cdot5\}$

Page 72   Exercise 7

1 $\{x: 1 \leqslant x \leqslant 4\}$ *2* $\{x: -1 \leqslant x < 3\}$ *3* $\{x: 0 < x < 5\}$

4 $\{x: x < 1\}$ *5* $\{x: x \geqslant -2\}$ *6* $\{x: x < -2 \text{ or } x \geqslant 2\}$

7 ———o——→
        3

8 ——●——
     1

9 ←———●———
         −2

10 ——o——→
      −1

11 ——●——●——
      −1   3

12 ←——o———●——→
        −2   2

13 ——o———o——
     −2   4

14 ——●————●——
      0   5

15 ←——●———o——→
       0   3

16 ←——————o——
             4

17 ——o——————→
     −1

18 ——————●——
            1

19 ——o———o——
     −3   3

20 ←————————●——
              3·5

21 ——o——————→
      3

22 ←————————●——
                4

23 ——o————·——→
     −2

24 —————————●——→
              2

25 ——o——————→
     −3

     ←——●——
       −1

     ←——————●——→
       −1   2

     ←—————o——
          −3
          —o——————
           −3   2

$\{x: -3 < x < 2, x \in R\}$

Page 74   Exercise 8

1 a $\{-2, 2\}$ *b* $\{x: -2 < x < 2\}$ *c* $\{x: x < -2 \text{ or } x > 2\}$

2 a $\{-2, 2\}$ *b* $\{x: -2 \leqslant x \leqslant 2\}$ *c* $\{x: x < -2 \text{ or } x > 2\}$

3 a ϕ *b* R *c* ϕ

4 a $-1, 3; -3$ *c* (*1*) $\{x: -1 < x < 3\}$ (*2*) $\{x: x < -1 \text{ or } x > 3\}$

5 a $-2, 4; 8$ *b* (*1*) $\{x: -2 \leqslant x \leqslant 4\}$ (*2*) $\{x: x < -2 \text{ or } x > 4\}$

6 $0, 4; 4$ *a* $\{x: 0 < x < 4\}$ *b* $\{x: x \leqslant 0 \text{ or } x \geqslant 4\}$

7 a $\{-2, 3\}$ *b* (*1*) $\{x: x < -2 \text{ or } x > 3\}$ (*2*) $\{x: -2 < x < 3\}$

8 $\{x: 1 < x < 3\}$ *9* $\{x: x < -5 \text{ or } x > -1\}$ *10* $\{x: -\frac{2}{3} \leqslant x \leqslant 1\}$

11 $\{x: -3 \leqslant x \leqslant 3\}$ *12* $\{x: x < -4 \text{ or } x > 3\}$ *13* $\{x: -1 < x < 2\}$

14 $\{x: x \leqslant -5 \text{ or } x \geqslant -\frac{1}{2}\}$ *15* $\{x: -2\frac{1}{2} < x < 3\}$

17 $\{x: x < 2 \text{ or } x > 2\}$ *18* ϕ *19* $\{x: -\frac{2}{3} < x < 4\}$

20a 0 or 6 *b* maximum $h = 45$ when $t = 3$ *c* $\{t: 0·8 < t < 5·2\}$

Algebra—Answers to Revision Exercises

Page 78   Revision 1A

1 $2(x^2 + 3)$ *2* $ab(bx + ay)$ *3* $p(p + q + r)$

4 $a^2(a - 1)$ *5* $(g - h)(g + h)$ *6* $(a - 1)(a + 1)$

7 $(b - 10)(b + 10)$ *8* $2(x - 5)(x + 5)$ *9* $(3a - 5b)(3a + 5b)$

10 $3(1-2y)(1+2y)$ 11 $x(x-1)(x+1)$ 12 $(1-7z)(1+7z)$

13 $(x-1)(x+2)$ 14 $(a+2)(a+13)$ 15 $(p-1)(p+5)$

16 $(y-6)(y-6)$ 17 $(2x+3)(2x+5)$ 18 $(2a-1)(a-4)$

19 $(3x+2)(2x+3)$ 20 $(6x+7)(x+1)$ 21 $(3x-5)(2x-3)$

22 $(3b+1)(b-1)$ 23 $(5a-1)(a-3)$ 24 $(2x+1)(x-3)$

25a 2   b   12   c   20   26a 100   b   22 000   c   53·4   27   1000

28 $\dfrac{x-y}{5}$ 29 $\dfrac{a-9}{2}$ 30 $\dfrac{n}{m+n}$ 31 $\dfrac{1}{3(x-3)}$

32 1   33 $\dfrac{2u+1}{u+3}$ 34 $\dfrac{2x+3}{(x+1)(x+2)}$ 35 $\dfrac{1}{6}$

36 $-\dfrac{2}{(x+1)(x-1)}$ 37 $\dfrac{u-v}{uv}$ 38 $\dfrac{ay+bx}{xy}$ 39 $\dfrac{9}{(x-5)(x-2)}$

40   a $P = 2r(2+\pi)$ b $A = r^2(4+\pi)$

Page 79   Revision Exercise 1B

1   $3x(x-2y)$ 2 $x^2(1-x)(1+x)(1+x^2)$

3   $t(t+8)(t-3)$ 4 $(a-b-10)(a-b+10)$

5   $(2p+2q-3r)(2p+2q+3r)$ 6 $(x-a+b)(x+a-b)$

7   $(x+6y)(x-4y)$ 8 $(4a+3b)(a-b)$

9   $(3m+8)(m+1)$ 10 $2(3p+1)(7p-5)$

11   $\left(x-\dfrac{2}{x}\right)\left(x+\dfrac{1}{x}\right)$ 12 $\left(t+\dfrac{1}{T}\right)\left(t+\dfrac{1}{T}-1\right)$

13   $(x-1)(x-y)$ 14 $(3a+b)(2x+3y)$

15   $(1-2m)(1+2m)(1+4m^2)$ 16 $(1-p+q)(1+p-q)$

17   $(x+6)(x-2)$ 18 $(c-a-b)(a+b+c)$

19   2·4   20   0·2   21 $t = u+4v$ 22 $2(a+b+c)$

23   $x-y = 5;\ x = 7,\ y = 2$

24   Factorise left side and substitute for x and y.

25· $\dfrac{x+y}{6x+y}$ 26   1   27 $\dfrac{2}{x}$

28   1   29 $\dfrac{x}{(x-2)(x+2)}$ 30 $\dfrac{1-ab}{a(a-b)}$

31   $\dfrac{1}{a(a-3)}$ 32 $\dfrac{1}{ab}$ 33 $\dfrac{8(x+1)}{(x-1)(x+3)}$

34 $\dfrac{5}{(x+1)(x-2)(x+3)}$ 35 $\dfrac{1}{a-1}$

36 $A = 2\pi r(h+2r)$, $V = \tfrac{1}{3}\pi r^2(3h+4r)$; 314 m^2, 360 m^3

Page 80 Revision Exercise 2

1 a $s = kt^2$ b $p = \dfrac{k}{d}$ c $I = ka^4$ d $P = \dfrac{k}{\sqrt{x}}$

2 a 12·8 b 2·5 3 a 3·2 b 1·5 4 80

5 a $F \propto x,\ Q \propto \dfrac{1}{v}$ b $F = 0·4x,\ Q = \dfrac{60}{v}$ 6 a $Q = \dfrac{70x}{3z}$ b 20

7 a $E = \dfrac{f^2}{6400}$ b $\dfrac{1}{25}$ 8 (6, 4), $(-12, -8)$, (15, 10), $(-3, -2)$, (5·7, 3·8)

9 a 1·67 m/s^2 b 3840 newtons 10a 200 ohms b 2 mm

11a $R = 1·3\ V + 8$ b 112 12b $v = 24 - 0·8t$

13b 1·13 c 53·1° 14a $R = 5 + 25V^2$ b 405

Page 83 Revision Exercise 3A

1 a $\{-5, 2\}$ b $\{-1, 1\}$ c $\{\tfrac{1}{2}, 10\}$
  d $\{0, 7\}$ e $\{0, -10\}$ f $\{8\}$

2 a $\{-4, -2\}$ b $\{2, 10\}$ c $\{-\tfrac{7}{2}, \tfrac{7}{2}\}$
  d $\{-2, \tfrac{1}{3}\}$ e $\{0, 8\}$ f $\{-13, 5\}$

3 $c = -36$; $\{-\tfrac{9}{2}, 4\}$ 4 7, (-9)

5 a (2, 0), (8, 0) b 0 and 10 c -9

6 a $\{-5, 15\}$ b $\{-\tfrac{7}{2}, \tfrac{5}{2}\}$ c $\{\tfrac{13}{4}, \tfrac{15}{4}\}$ d $\{-\tfrac{3}{20}, \tfrac{13}{20}\}$ 7 13

8 a $\{-4·4, 1·4\}$ b $\{-0·8, ·2·4\}$ c $\{-1·1, 0·7\}$
  d $\{-1·4, 1·0\}$ e $\{-0·3, 1·8\}$ f $\{-2·8, 0·8\}$ 9 5

10 a $\{x: x < -3\ \text{or}\ x > 3\}$ b $\{x: 5 \leqslant x \leqslant 8\}$
  c $\{x: x \leqslant 3\ \text{or}\ x \geqslant 4\}$ d $\{x: -\tfrac{1}{2} < x < 3\}$

12 32 13 a $\{-2, 3\}$ b (1) $\{x: x < -2\ \text{or}\ x > 3\}$ (2) $\{x: -2 < x < 3\}$

14 5; 30 cm^2 15 3 or -1

Page 84 Revision Exercise 3B

1 a $\{-\tfrac{9}{2}, \tfrac{3}{2}\}$ b $\{-\tfrac{7}{4}, 2\}$ c $\{-\tfrac{1}{3}\}$ d $\{-\tfrac{1}{2}, 0\}$

2 a $\{-\frac{3}{4}, \frac{7}{2}\}$ b $\{-7, 1\}$ c $\left\{-\dfrac{1}{\sqrt{2}}, \dfrac{1}{\sqrt{2}}\right\}$ d $\{2, 2\frac{1}{2}\}$

3 a $v = \pm\sqrt{\dfrac{2eV}{m}}$ b $r = \pm\sqrt{\dfrac{q}{Ek}}$ c $r = \pm\sqrt{\dfrac{A}{4\pi}}$ d $f = \pm\dfrac{1}{2\pi\sqrt{LC}}$

4    $\frac{3}{4}$ or $-\frac{2}{3}$ 5 a $(-3, 0), (8, 0)$ b $(0, -6), (0, 4)$ 6 a   1 or 5   b   3

7 a $\{-3, 6\}$ b $\{-10, 5\}$ c $\{-\sqrt{8}, \sqrt{8}\}$ d $\{\frac{1}{2}\}$ (-1 not a solution)

8   1 or 4; after 1 second on the way up, and after 4 seconds on the way down.

9    $\{-8, 1\}$ (1) $\{x: x < -8 \text{ or } x > 1\}$ (2) $\{x: -8 < x < 1\}$

10a $\{-1.3, 0.9\}$ b $\{-0.7, 0.9\}$ c $\{0.2, -1.0\}$

12   $\{x: -10 < x < 4\}$ $\{x: x \le -1 \text{ or } x \ge 3\}$ $\{x: x < \frac{1}{2} \text{ or } x > \frac{1}{2}\}$

a ○———○   b ←●———●→   c ←○———→
  −10   4   −1   3 $\frac{1}{2}$

Algebra—Answers to Cumulative Revision Exercises

Page 86   Cumulative Revision Exercise A

1 a F b T c F d T e F f T g F

2 a $\{3, 4, 5, 6, 7\}$ b $\{2, 4, 6, 8, 10\}$ c $\{1, 2, 3\}$

3 a $4p^2 - 3q^2$ b   6b   c $3x^2 - 4$ d $12x + 24$, or $12(x + 2)$

4 a 0 b 10 c 23 d 1

5

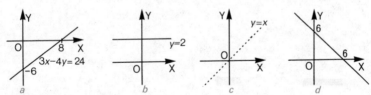

6 a 6 b 26 c $p^2 + q$ d 36 e 36 f $(p + q)^2$

7 q $\{-1\}$ b $\{x: x > 3\}$ c $\{10\frac{1}{2}\}$ d $\{0\}$ e $\{3\}$

8   (i) $A \cup B \cup C$ (ii) $A \cup (B \cap C)$ (iii) $(A \cap B) \cup C$

9 a 2.4 or 0.6 b 1.9 or -0.9 c -0.18 or -0.42 d -5.35 or 1.35

10a $\{d, e\}$ b $\{a, b, c, d, e, f\}$ c $\{f, g\}$ d $\{a, b, c, f, g\}$ e $\{a, b, c, f, g\}$

11   a and d   12a 12 b -42 c -10 d 0

13a $\{(4, 1)\}$ b $\{(2, -1)\}$ c $\{(-1, 0)\}$

14a T b F c F d F e T

15a $a(b-c)(b+c)$ *b* $p(x^2-y^2+z^2)$ *c* $(2x-1)(2x+1)$ *d* $(2x-3y)(2x+3y)$

e $(x-4)(x+3)$ *f* $(2x-1)(x-2)$ *g* $5pq(p-q)$ *h* $(3a-7)(a-1)$

16a 28   *b* 2 and -2 *17a* a^2+4a-5 *b* $6x^2-13x+6$

c p^2-q^2 *d* $b^2+8b+16$ *e* $4c^2-20c+25$ *f* $25x^2-40xy+16y^2$

18 5   *19a* A(0, 4), B(4, 4), C(8, 0); (0, 8)

b $\{(x, y): x\geqslant 0,\ y\geqslant 0,\ y\leqslant 4,\ y\leqslant -x+8;\ x,\ y\in R\}$

20a $-2{\cdot}25$ *b* $x=1{\cdot}5$ *c* 0 and 3   *d* $\{x: 0<x<3, x\in R\}$ *e* $k=4$

f $\{(x, y): y\leqslant x,\ y\geqslant x^2-3x;\ x,\ y\in R\}$

21a $\dfrac{x-3}{2}$ *b* $\dfrac{1}{x+2}$ *c* $\dfrac{x-1}{3}$ *d* $\dfrac{x-3}{x-2}$

22a $\{(1, 3), (1, 5), (1, 7), (3, 5), (3, 7), (5, 7)\}$ *b* $\{(1, 1), (3, 3), (5, 5), (7, 7)\}$

c $\{(3, 1), (5, 1), (7, 1), (5, 3), (7, 3), (7, 5)\}$

Domains and ranges: *a* $\{1, 3, 5\}, \{3, 5, 7\}$ *b* $\{1, 3, 5, 7\}, \{1, 3, 5, 7\}$

c $\{3, 5, 7\}, \{1, 3, 5\}$

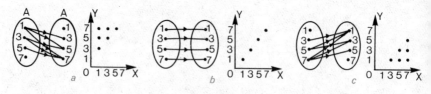

23a (*1*) 4   (*2*) 2 or -2 *b* (*1*) 5   (*2*) 4   *24a* 8·5   *b* 3

25a $\{-2, \frac{1}{2}\}$ *b* $\{x: -2<x<\frac{1}{2}, x\in R\}$ *c* $\{x: x<-2 \text{ or } x>\frac{1}{2}, x\in R\}$

26a $\dfrac{C}{\pi}$ *b* $\dfrac{A}{2\pi r}$ *c* $\sqrt{\dfrac{V}{4\pi}}$ *d* $\dfrac{y-q}{p}$ *e* $\dfrac{P-2b}{2}$ *f* $\left(\dfrac{ky}{z}\right)^2$

27

E A ──── S ──── M
(66−x) (x) (12) 24 (72−x)

$x = 14$

28a $\dfrac{2a-1}{a(a-1)}$ *b* $\dfrac{u}{(u-3)(u-2)}$ *c* $\dfrac{2x}{(2x-1)(2x+1)}$

29a $\{-4, 2\}$ *b* $\{-2{\cdot}5, -1\}$ *c* $\{-2{\cdot}8, -0{\cdot}7\}$ *30* $\frac{3}{20}$

31a $-4, -\frac{32}{9}, 2$ *b* 5 and -5 *32* 17·3

33a $\{x: x \geqslant 1\}$ *b* $\{2, 3\}$ *34* $-6 = 2m+c, 9 = 3m+c; m = 15, c = -36$

35b (1) and (2) $\{2\cdot4, -0\cdot4\}$ *36* $a = 2, b = 12$

37a $2 \times 3, 2 \times 3, -2, -1$

b (1) $\begin{pmatrix} 3 & -3 & 2 \\ -2 & 2 & 9 \end{pmatrix}$ (2) $\begin{pmatrix} 3 & -1 & 0 \\ 2 & -4 & -1 \end{pmatrix}$ (3) $\begin{pmatrix} 9 & -9 & 6 \\ -6 & 6 & 27 \end{pmatrix}$.

38a $x = -2. y = -2$ *b* $x = -4, y = 0$

39a (-14) *b* $(ax+by+cz)$ *c* $\begin{pmatrix} 6 & -4 \\ -1 & 1 \end{pmatrix}$ *40* $(2, -1), (0, -3), (-4, 2)$

41a $\begin{pmatrix} 3 & -2 \\ -4 & 3 \end{pmatrix}$ *b* $\begin{pmatrix} 1 & 0 \\ 0 & 1 \end{pmatrix}$ *c* $\frac{1}{4}\begin{pmatrix} 2 & 0 \\ 0 & 2 \end{pmatrix}$ *d* $\frac{1}{2}\begin{pmatrix} 5 & -3 \\ 4 & -2 \end{pmatrix}$

42a $\{(2, 1)\}$ *b* $\{(1, -2)\}$ *c* $\{(3, 0)\}$

Page 91   Cumulative Revision Exercise B

1 a $\{16\}$ *b* $\{x: x \geqslant 4\}$ *c* $\{x: x < \frac{1}{2}\}$ *d* $\{x: x < 3\frac{3}{4}\}$ *e* $\{-\frac{2}{3}\}$

2 a 210 *b* 6   *3 a* $12x^2 + 18xy - 12y^2$ *b* $x^4 - 1$ *c* $y^6 + 2y^3 + 1$

d $x^2 + 2 + \dfrac{1}{x^2}$ *e* $x^3 + y^3$ *f* $x^2 + 1 + \dfrac{1}{x^2}$

4 a $x^2 = 25 \Rightarrow x = 5$; T, F *b* $a = 0$ and $b = 0 \Rightarrow ab = 0$; F, T

c $q < p \Rightarrow p > q$; T, T   *d* $\cos a° = 0 \Rightarrow \sin a° = 0$; F, F

e If in triangle ABC, $a^2 = b^2 + c^2$, then the triangle is right-angled at A; T, T

5 a $(a-b)(a+b)(a^2+b^2)$ *b* $px(x-1)(x+1)$ *c* $(p-5q)(p+4q)$

d $2x^2(x-1)(x+1)(x^2+1)$ *e* $2(2a+1)(a-3)$ *f* $(2a-3b-1)(2a-3b+1)$

g $-5(a+b)(a-b)$ *h* $a^2x^2(x+a-1)$ *i* $(\cos A - 1)(\cos A + 3)$

6 $P = 2x(\pi+1), A = x^2(\pi+2)$; 41·4 m, 128 m^2

7 a $x = \dfrac{z}{y+z}$ *b* $x = \dfrac{b}{\sqrt{(1+a^2)}}$ *c* $x = \dfrac{ab+cd}{b}$ *d* $x = \dfrac{yz}{y-z}$

8 a $-5, 1, -1$ *b* 0 and -1 *9* $a = -1, b = -2$

10a $\dfrac{-3x}{(x+1)(2x-1)}$ *b* $\dfrac{-a}{2(a+2)^2}$ *11* $K\left(\dfrac{c}{a-m}, \dfrac{ac}{a-m}\right)$

12b reflection in *y*-axis; reflection in *x*-axis *c* either reflection followed by the other gives a half turn about the origin.

Geometry—Answers to Chapter 1

Page 98   Exercise 1

1 a coincide with original at regular intervals   *b* coincides all the time

2   same   *3* (i) 180°   (ii) 120°, 240°   (iii) 90°, 180°, 270°

(iv) 45°, 90°, 135°, 180°, 225°, 270°, 315°

4   (i) 2   (ii) 3   (iv) 8   *5* any number, or an infinite number

Page 99   Exercise 2

1   In Fig. 4 make ∠AOA' = 30°, 60°

2 a anticlockwise, or positive

b OA = OA', AB = A'B', OB = OB'; ∠OAB = ∠OA'B', ∠ABO = ∠A'B'O. ∠AOB = ∠A'OB'

c congruent   *d* 90°   *e* O   *f*

3 a OA', OB'   *b* 40°   *c* 40°

4 a OA', OB'   *b* 140°   *c* 40°

5

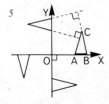

6 a (−5, 5)   *b* (5, −5)   *c* (−5, −5)

d (0, 5√2)   *e* (5√2, 0)

7 a C   *b* D   *8* R, S, M; same answer in each direction

9   180° in each case   *10a* R   *b* E   *c* P   *d* OB   *e* TU

11a 60°, 120°, 180°, 240°, 300°   *b* 6

12a should coincide after rotation of 90°

b should coincide after any rotation   *c* try to fit after rotation

Page 102   Exercise 3

1 a AB   *b* OA, OB, OR, OS   *c* PQ   *d* PJQ   *e* PNQ or RS   *f* ORS

2 a yes, no   *b* (2)

3 a equilateral *b* isosceles *c* isosceles; possibly; no

4 a at opposite ends of a diameter 5

b diameter, circumference, centre, radius

Page 103 Exercise 4

1 a,b,c equal in each case *2 a,b,c* equal in each case *d* 1:2

3 c $x°, 2x°, 5x°$ *d* $y, 2y, 5y$ cm *e* $z, 2z, 5z$ cm^2 *f* 1:2:5 in each case

4 a 1:2 *b* 1:3 *c* 2:3 *d* 1:2 *e* 1:3 *f* 1:12 *g* 1:2 *h* 5:12

5 8 cm, 12 cm, 48 cm *6* 72 cm^2 *7 a* PQ, RS, TU *b* 6 m *c* 10 cm^2

8 a T *b* F *c* T *d* T *e* F *9 a* 1:3 *b* $\frac{1}{3}$; 4 cm *c* $\frac{1}{4}$; 216 cm^2

10a 18 cm *b* 28° *c* $x:y = a:p$

Page 107 Exercise 5

1 a ∠POQ *b* ∠SOR *c* vertically opposite, equal

d ∠POS, ∠QOR, vertically opposite, equal

2 ∠AOB, ∠COD; chords are equal

3 Equal chords are equidistant from the centre.

4 a OM *b* ∠XOZ *c* XZ *d* OYX

5 a 6 cm, 5 cm *b* both perpendicular to PQ

6 Equidistant chords are equal.

7 Equal chords are equidistant from the centre.

Page 109 Exercise 5B

1 Equal chords are equidistant from the centre.

2 a T *b* F *c* F *d* T

3 a 135° *b* $\frac{2}{15}$ *c* $22\frac{1}{2}°$ *4* △AOB is equilateral

5 a 36°, 72°, 108°, 144°, 180°, 216°, 252°, 288°, 324°

b −108°, −144°, −180°, −216°, −252° *c* 144°

6 a 0°, 120°, 240° *b* A, OB, COB *c* 6 equal sides, 6 equal angles

Page 110   Exercise 6

1 a They fit their outlines after rotations of 120°, 90°, 72°, 60° respectively.
 c 3, 4, 5, 6 *d* equilateral △, square

2   join ends of radii   *3 a*   120°, 90°, 72°, 60°

4   60°, 60°, 120°, 6; 30°, 75°, 150°, 12; 3·6°, 88·2°, 176·4°, 100;

$$\frac{360°}{n}, 90° - \frac{180°}{n}, 180° - \frac{360°}{n}, n$$

Geometry—Answers to Chapter 2

Page 113   Exercise 1

1 a 5 *b* 5 *c* 5   *2 a* 4, 4, 4 *b* 6, 6, 6

Page 115   Exercise 2

1 b perpendicular to PQ

2 b ST is the perpendicular bisector of MN

3 a XY *b*

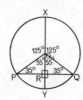

c PR = RQ = 5 cm

d OPR, OQR

e PY = YQ, PX = XQ

4 a XY *b* CN = ND, AM = MB *c* CX = XD, AC = BD, AY = YB

5 kite *6 a* ST *b* SP, SQ; TP, TQ *c* SP, SQ; TP, TQ

 d ∠STX, ∠STY, ∠SKP, ∠SKQ, etc.

7 b BD, arc AC, ∠BDC *c* ∠BCD, ∠DAB, ∠CBA

Page 116   Exercise 2B

1   chord perpendicular to OA

2 a AB = CD *b* AB < CD *c* AB > CD

3 a OM = ON *b* OM < ON *c* OM > ON

4 a nearer the centre *b* greater

Page 117 Exercise 3

3 a perpendicular bisector of AB

 b any number; perpendicular bisector of AB

 c yes; no *d* for $r<\frac{1}{2}$AB, none; $r=\frac{1}{2}$AB, one; $r>\frac{1}{2}$AB, two

4 a on perpendicular bisector of AB *b* on perpendicular bisector of BC

 d yes; one *5* perpendicular bisectors are parallel

6 d inside the triangle, outside the triangle, midpoint of hypotenuse

7 OA = OC. They are concurrent. *8 c* no *d* yes

9 a usually impossible *b* always possible *c* usually impossible

 d always possible *e* possible only when kite has two right angles

Page 119 Exercise 4

1 a 6 cm *b* 16 cm *c* 128 cm^2 *2 a* 7 m *b* 32 m *c* 768 m^2

3 a 15 cm *b* 30 cm *c* 30 cm *4 a* 12 m *b* 9 m *c* 18 m

5 1·2 m, 0·8 m *6* 13 cm, 67·4° *7* 19·1 cm, 18·3 cm

8 a 1·6 cm *b* a concentric circle with radius 1·6 cm

Page 120 Exercise 4B

1 14 cm, 2 cm *2* 15 m *3* 80 cm *4* yes, 72 cm

5 a 20, 19·9, 19·6, 19·1, 18·3, 17·3, 16·0, 14·3, 12·0, 8·72, 0

 c 15·2 cm, 6·62 cm

Page 121 Exercise 5

1 a Pythagoras' theorem *b* $e^2+f^2=36, g^2+h^2=36$ *2* $x^2+y^2=25$

3 a on the circumference *b* outside the circle *c* inside the circle

4 $x^2+y^2=r^2$

Page 122 Exercise 6

1 $x^2+y^2=4, x^2+y^2=25, x^2+y^2=36, x^2+y^2=100, x^2+y^2=144$

2 a 2 *b* 7 *c* 1 *d* 20

3 a $x^2+y^2=9$ *b* $x^2+y^2>100$ *c* $x^2+y^2<64$

4 a O, 4 *b* O, 10 *c* O, a *d* O, 0·5 *5 d*

6 a on the circumference *b* outside the circle *c* inside the circle

7 a $x^2 + y^2 = 81$ *b* $x^2 + y^2 > 64$ *c* $x^2 + y^2 < 36$

8

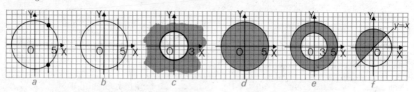

a   b   c   d   e   f

9   It is the empty set.

10a 5, $x^2 + y^2 = 25$ *b* $-4, -3$ *c* $(-3, -4)$

  d $\pm 3, \pm 3, \pm 4, \pm 4$ *e* yes; 90°, 180°, 270°

11a ± 1 *b* ± 5 *c* ± 3 *d* ± 6

12a $(\pm 4, 3)$, 8   *b* $(\pm 3, 4)$, 6

  c $(\pm 0.995, 4.9)$, 1.99   *d* (0, 5), 0 *e* no intersection

Page 124   Exercise 6B

3 $x^2 + y^2 = 4$; 2   *4* $x^2 + y^2 = 9$; 3

5 $x^2 + y^2 = 16$; 4   *6* $x^2 + y^2 = 25$; 5   *7* $x^2 + y^2 = 18$; $\sqrt{18}$

Geometry—Answers to Chapter 3

Page 127   Exercise 1

1 a corresponding angles with parallel lines

  b $\angle OP_1 Q_1 + \angle OP_1 U_1 = \angle OP_2 Q_2 + \angle OP_2 U_2$, etc.

  c $\frac{3}{2}$ *d* equiangular and sides in proportion

2 e $OP_1 Q_1, OP_2 Q_2$; $OQ_1 R_1, OQ_2 R_2$; etc.

  f $P_1 Q_1 \| P_2 Q_2$; $Q_1 R_1 \| Q_2 R_2$; etc.

4   They are the same; the enlargements are congruent.

5   *6*   *7*   *8*

9

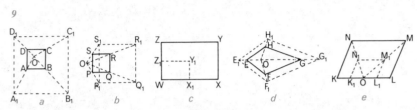

10 (i) 2, enlargement   (ii) $\frac{1}{2}$, reduction

(iii) 3, enlargement   (iv) $\frac{1}{3}$, reduction

11 (i) $\frac{1}{4}$ (ii) 3 (iii) 3

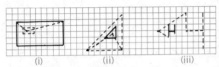

12c lengths in $A_1 B_1 C_1$ $\frac{1}{4}$ of those in $A_2 B_2 C_2$; corresponding sides parallel

Page 131 Exercise 2

1 (i) $\overrightarrow{OP'} = 5\overrightarrow{OP}$ (ii) $\overrightarrow{OP'} = \frac{3}{2}\overrightarrow{OP}$ (iii) $\overrightarrow{OP'} = \frac{1}{3}\overrightarrow{OP}$

2

3 $(-1, 3), (3, 3), (3, 4), (-1, 4)$ 4 $(-3, 9), (9, 9), (9, 12), (-3, 12)$

5 $(8, 0), (0, -4); (1, 0), (0, -\frac{1}{2})$

6 d $(1, 1), (3, 2), (2, 4); (\frac{5}{2}, \frac{5}{2}), (\frac{15}{2}, 5), (5, 10)$ e multiplied by $\frac{1}{2}$ and $\frac{5}{4}$

7 d $(3, 3), (9, 6), (6, 12); (\frac{3}{2}, \frac{3}{2}), (\frac{9}{2}, 3), (3, 6)$ e $[O, \frac{3}{4}]$

8 $\overrightarrow{OB'} = 4\overrightarrow{OB}; 4\overrightarrow{OB} - 4\overrightarrow{OA} = 4(\overrightarrow{OB} - \overrightarrow{OA}) = 4\overrightarrow{AB}; A'B' \| AB, A'B' = 4AB$

11a 2 b 3 c $\frac{4}{3}$ d $\frac{4}{3}$ e 2 f $\frac{3}{2}$ g $\frac{3}{2}$ h $\frac{6}{5}$

Page 133 Exercise 3

1

2 a -1 b (1) false (2) true (3) true

3 -2

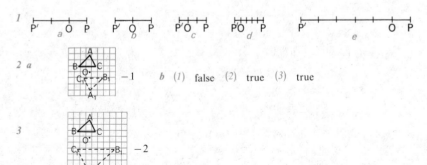

4 $(-6, 0)$, $(-9, -3)$, $(-3, -3)$ *5* $(-2, 0)$, $(-3, -1)$, $(-1, -1)$

7 b $(5, 0)$, $(2, 0)$, $(-3, -4)$ *c* a half turn about O, or reflection in O

8 c $(2, 2)$, $(-6, -2)$, $(-10, -10)$, $(-2, -6)$

9 $(11, -3)$, $(-1, -3)$, $(-1, -9)$, $(11, -9)$; the centre is not at O

Page 136 Exercise 4

2 a (*1*) yes (*2*) no (*3*) no

b (*2*) line parallel to axis (*3*) half turn

3 a on the line *b* at the point *c* at the centre

4

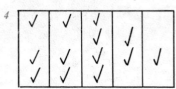

5 a $(20, 10)$, $(-10, 15)$, $(-5, 0)$, $(-10, -5)$

b $(-12, -6)$, $(6, -9)$, $(3, 0)$, $(6, 3)$

c $(-6, -3)$, $(3, -\frac{9}{2})$, $(\frac{3}{2}, 0)$, $(3, \frac{3}{2})$

6 images $(-2, -4)$, $(-4, -4)$, $(-4, 1)$, $(-2, 1)$

7 c (*1*) $(8, 8)$, $(4, 4)$, $(2, 2)$, $(1, 1)$, $(\frac{1}{2}, \frac{1}{2})$, $(\frac{1}{4}, \frac{1}{4})$, $(\frac{1}{8}, \frac{1}{8})$

(*2*) $64, 16, 4, 1, \frac{1}{4}, \frac{1}{16}, \frac{1}{64}$ square units.

8 b they are similar; their sides are parallel to those in the other rectangles.

9 a $AB = \sqrt{13}$, $BC = \sqrt{52}$, $CA = \sqrt{65}$, so $CA^2 = AB^2 + BC^2$

b $(8, 2)$, $(14, 6)$, $(6, 18)$

c $A'B' = \sqrt{52}$, $B'C' = \sqrt{208}$, $C'A' = \sqrt{260}$, so $(C'A')^2 = (A'B')^2 + (B'C')^2$

10a $(-1, -2)$, $(0, 0)$, $(1, 2)$, $(2, 4)$, etc.

b $(-3, -6)$, $(0, 0)$, $(3, 6)$, $(6, 12)$, etc. *c* $y = 2x$

11a $(-1, 4)$, $(0, 0)$, $(1, -4)$, etc.

b $(-3, 12)$, $(0, 0)$, $(3, -12)$, etc. *c* $y = -4x$

12a $(3, -1)$, $(3, 0)$, $(3, 1)$. etc. *b* $(9, -3)$, $(9, 0)$, $(9, 3)$, etc. *c* $x = 9$

13a $(-1, 2)$, $(0, 4)$, $(1, 6)$, etc. *b* $(-3, 6)$, $(0, 12)$, $(3, 18)$, etc. *c* $y = 2x + 12$

14a $(-2, 4)$, $(0, 8)$, $(2, 12)$, etc.; $y = 2x + 8$

b $(\frac{1}{2}, -1)$, $(0, -2)$, $(-\frac{1}{2}, -3)$, etc.; $y = 2x - 2$

15 $y = mx$ *16* $y = 2x + 4k$ *17b* $x = 3$, $y = 6$, $x + y = 12$

Page 138   Exercise 4B

1 a   (0, 0), (±1, 1), (±2, 4), etc. *b*   (0, 0), (±2, 2), (±4, 8), etc.

2 a   (0, 0), (±4, 4), (±8, 16), etc.; $y = \frac{1}{4}x^2$

  b   (0, 0), (±2, −2), (±4, −8), etc.; $y = -\frac{1}{2}x^2$

3 $y = \frac{1}{k}x^2$ *4 a*   (11, 3), (13, 3), (13, 5), (11, 5)

  b   (− 2, −1), (2, −1), (2, 3), (−2, 3) *c*   (1, 1), (7, 1), (7, 7), (1, 7)

5 a   3, 6; 4, 8; 1, 2 *b*   −1, −2; 1, 2; −1, −2

  c   −2, −4; 1, 2; −1, −2; second difference is twice the first

6 b   (0, 5); (3, −1), (12, −4) *c*   (2, 4); (3, 5), (0, 6)

7 b   (1, 9); 2 *c*   (13, 7); −2

8 a $\overrightarrow{OC'} = \overrightarrow{OA'} + \overrightarrow{A'C'} = k\overrightarrow{OA} + k\overrightarrow{AC} = k(\overrightarrow{OA} + \overrightarrow{AC})$

  b   O, C, C' collinear: OC' = k OC

Geometry—Answers to Revision Exercises

Page 144   Revision Exercise 1

1 a   PQ *b*   OP or OQ *c*   PR *d*   PR or RQ *e*   RQP or QPR

2   (i)   2 (ii) ◂4 (iii)   3 (iv)   infinite

3   A'(−1, 3), B'(−6, 1), C'(−4, −2)   *4*   120°

5 a   (*1*)   C (*2*)   arc EF (*3*)   chord GH (*4*)   sector KOA (*5*)   △BOD

  b   (*1*)   equal (*2*)   each = 140°   *6 a*   24 cm *b*   5° *c*   1:18

7 a   6 cm *b*   9 cm²   *8 a*   15·7 cm *b*   141 cm²

9   112°   *10*   OP = OS, OQ = OT, OR = OU

11a   diameter parallel or perpendicular to PQ   *b*   diameter bisecting the two unequal arcs

12a   18° *b*   162°   *13*   36°   *14a*   85°, 85°, 10° *b*   36   *15*   9 cm²

Page 147   Revision Exercise 2

1   (i)   2 (ii)   4 (iii)   3 (iv)   an infinite number

2 a   straight line through two centres *b*   any line through common centre

3   Draw chord EPF, where E is the image of C in AB; PD = PF, EP = CP, CD = EF

4 a PQG *b* AB = DE, BC = EF, CG = FG, PM = PN, etc

5 a T *b* F *c* T *d* T *6 a* 1·2 cm *b* 3·2 cm *c* 5·12 cm²

7 a each = 3 cm *b* equal radii; centre O, radius 3 cm

8 a 5 cm, 11·5 cm, 12·6 cm *b* no

9

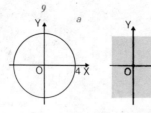

a

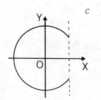

b

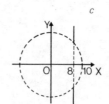

c

10

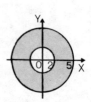

11a (±6, 8), 12 *b* (−8, ±6), 12

 c (±8, −6), 16 *d* (10, 0), 0 *e* no intersection

12 circle, centre O, radius = distance from O to one of the chords

13 9·5 cm *14a* 12 cm *b* 8·5 cm *15* 25 cm, 29 cm

16 25 cm *17* 50·5 m *18* 29 cm

19

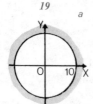

a

b

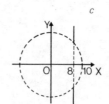

c

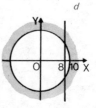

d

20 acute-angled, right-angled, obtuse-angled triangles

Page 150 Revision Exercise 3

1 b $A_1(6, 2)$, $B_1(0, 6)$, $C_1(−4, −4)$ *c* They are similar.

2 b (0, 0), (−3, −1½), (−3, −3), (−1½, −3) *c* They are similar.

3 b [O, 3] *c* [O, ⅓] *4 a* [O, 3] *b* [O, ⅓] *c* centre O, radius 2

5 a (9, 6) *b* (−3, 12) *c* (6, −15) *d* (0, 0)

6 a (−6, −4) *b* (2, −8) *c* (−4, 10) *d* (0, 0)

7 a F *b* T *c* (*1*) T (*2*) F (*3*) F *d* T *e* T

8 C, A, B *9 a* (−1, −2), (0, 1), (1, 4), etc.

 b (−2, −4), (0, 2), (2, 8), etc. *c* $y = 3x + 2$

10b ⅖ *c* −6 *d* 1 *11b* (3, 3), 3 *c* (8, 6), 6 *d* (−⅖, 0), 6

12a $2a, 2b, 2c, 2d$ *b* $AC \| A_1 C_1 ; A_1 C_1 = 2AC$

c $-a, -b, -c, -d$; $\overrightarrow{A_1 C_1}$ and $\overrightarrow{AC}$ are equal in magnitude but have opposite directions.

13 (9, 4), 3

Geometry—Answers to Cumulative Revision Exercise

Page 153 Cumulative Revision Exercise

2   22 or 23 m   *3 a* 6, 8, 12 *b* 4, 4, 6 *c* 5, 5, 8 *d* 7, 10, 15

4   (8, 1), (8, 5), (2, 5)   *5 a* isosceles *b* right-angled isosceles

6   7·7 km   *7*   *8*

9 a rectangle, two parallelograms, two isosceles triangles, kite

b three parallelograms, three kites

10   0·9 m   *11a* 10 *b* 13 *c* $\sqrt{50}$ or 7·07   *12a* 10 cm *b* 96 cm^2

13a $\overrightarrow{AC}$ *b* $\overrightarrow{AM}$ *c* $\overrightarrow{AD}$ *d* $\overrightarrow{MD}$ *14*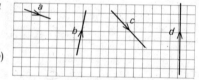

15a $\frac{1}{3}u$ *b* $\frac{2}{3}u$ *c* $\frac{2}{3}u$ *d* $-\frac{1}{3}u$ *e* $-\frac{2}{3}u$

16 $b-a$ *a* $\frac{2}{3}(b-a)$ *b* $\frac{1}{3}(b-a)$ *c* $\frac{1}{3}(a+2b)$

17a $u+v$ *b* $\frac{1}{2}(u+3v)$

18a $\begin{pmatrix} 3 \\ 2 \end{pmatrix}$ *b* $\begin{pmatrix} 6 \\ -9 \end{pmatrix}$ *c* $\begin{pmatrix} -8 \\ 0 \end{pmatrix}$ *d* $\begin{pmatrix} 5 \\ -1 \end{pmatrix}$ *e* $\begin{pmatrix} 11 \\ 2 \end{pmatrix}$ *f* $\begin{pmatrix} -8 \\ -6 \end{pmatrix}$; $\sqrt{13}, \sqrt{13}, 4$

19   equiangular; AB, DE; BC, EF; CA, FD

20   sides proportional; A, P; B, Q; C, R

21a yes; equiangular, sides proportional *b* no; sides not proportional

22   9 cm   *23*   20 mm

24a *(1)* -1 *(2)* cannot be found *(3)* 0 *b* $y = \frac{1}{2}x+3$ *c* $-\frac{4}{3}$, (0, -2)

25a ∠TPQ   *b* ∠QOR and ∠QPR

26a 24 cm   *b* 20 cm²   *27* 8·37 cm

28a 13 cm   *b* 531 cm²; obtuse-angled △

29a 17·3 cm   *b* circle, centre P, radius 5 cm

30 make angles of 72° at centre   *31* 8·5 cm

32a 3, 3   *b* 4, 4   *c* 6, 6   *d* 8, 8; each = number of sides; *n, n*

33a $x^2 + y^2 = 9$, $x^2 + y^2 = 25$, $x^2 + y^2 = 81$ *b* 1   *c* $x^2 + y^2 = 18$

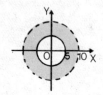

34a points on or outside circle, centre O, radius 5

  b points inside circle, centre O, radius 10

35 2; $p = 6$, $r = 6$ *36* 2, (4, 4); −3, (1, 3)

37

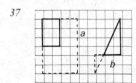

38a centre O, radius 6 cm

  b centre O, radius 2 cm

39a (6, 8), (10, −2), (−6, 0)   *b* (−9, −12), (−15, 3), (9, 0)

  c (3, 4), (5, −1), (−3, 0)   *d* $(\frac{3}{2}, 2)$, $(\frac{5}{2}, -\frac{1}{2})$, $(-\frac{3}{2}, 0)$

40a (−1, −3), (0, 1), (1, 5), etc.   *b* (−2, −6), (0, 2), (2, 10), etc.

  c $y = 4x + 2$

41 $\binom{4}{6}$, $\binom{-6}{4}$, $\binom{4}{6}$, $\binom{-6}{4}$; all same magnitude, AB‖DC, BC‖AD; square

42 $\sqrt{8}$

Arithmetic—Answers to Chapter 1

Page 164   Exercise 1

1 downwards, indicating the fall in the purchasing power of £1.

2 1920 and 1934   *3* The purchasing power fell sharply during the wars.

4 a 1918, 1922 and 1940   *b* 1952   *5* 84%

Page 165   Exercise 2

1 b £2366   *c* £146·64   *2* £2000   *3* £52·50

4 a upwards   *b* import

  c Both show increases, but wages have risen faster than prices.

  d large increases towards end of each year.   *e* 78–79%; 42–43%

5 a £4·10; £8·25   *b* 30%

6 a 1968   *b* food 1971; housing 1968   *c* 60; 75   *d* 14% approximately

Page 168   Exercise 3

1   71·70   *2* 823   *3* £3·89   *4* 17 100; £2·11

5   £20   *6 a* £70   *b* £155·20   *c* £207   *7* £81·56

8   £6·64   *9* £201   *10* £255·70

Page 171   Exercise 4

1   £2·60   *2* £7·55   *3* £7·22   *4* £32   *5* £10   *6* £6·25

7   £16·04   *8* £4·20   *9* £57·50; 16%

Page 174   Exercise 5

1   £15   *2* £14   *3 a* £10   *b* £525

4   £1025, £1061, £1124, £1187, £1250

5 a 82·8p   *b* 2·07%   *6* £195 000

7 a 5%   *b* 33   *c* £4·95   *8 a* £720, £840   *b* 5·7%

9   £1420   *10* £1625; £755   *11* £24·62   *12* £2006·4 million

13 3%, 4·4%, 7·0%, 8·7%

Page 177   Exercise 6

1   £40·80   *2* £14·14   *3* £55·35   *4* £48·96   *5* £122·58

6   £224   *7* £100·73   *8 a* £80   *b* £81·60 £1080; £1081·60

9   57·7 million, 59·4 million   *10* £506   *11* £512   *12* £280·35

13 £41, £15, £54, £51, £122, £222

310 ANSWERS

Page 180   Exercise 7

1 a £4·75   *b* £3·65   *c* £2·20   *d* £1·60

2 £69·30   *3* £79·86   *4* £39·60   *5* £790

6 With-profits premiums are about the same amount greater than without-profits premiums throughout; both types of premium increase with age, more steeply from about 40 onwards.

7 about £608   *8* £1026; £315

Page 182   Exercise 8

1 a £3   *b* £18·75   *c* £1·50   *d* £12·43

2 £15·50   *3* £1·25   *4 a* £14·25   *b* £20·90

5 £8·65   *6* £12·65   *7* £17·85

8 a £13·80   *b* £1·68   *c* 35p   *d* £112·50

Page 183   Exercise 9

1 a £27·50   *b* £30   *c* £54   *2* £63   *3* £25·20

Page 185   Exercise 10

1 £61·60   *2 a* £104   *b* £59·50   *c* £130·20   *d* £118·80

3 £16·20   *4* £15·60 more   *5* £40   *6* 46%   *7* £2·69   *8* £3·66

Page 186   Exercise 11

1 a 80p   *b* £1·05   *c* 76p   *2 a* £80   *b* £105   *c* £76

3 a 76p   *b* £89·68   *4* 79p   *5* 87p; £14000

Page 188   Exercise 12

1 environmental services; social services; defence; trade etc.; income tax, customs and excise, tax on companies, purchase tax

2 4% *3* 39% *4* £62·5 million

Page 190 Exercise 13

1 a £300 *b* £384 *c* £190·50 *d* £70 *2* £55·75 *3* £1251

4 a £105 *b* £73·50 *5* £32·30 *6* £35·18 *7* £42·93

8 a £2850 *b* £579·16

Page 192 Exercise 14

1 a £57·75 *b* £315·92 *c* £9·15 *d* 93·5p *e* £30·91 *f* £28·53

2 a £5·10 *b* £76·40 *c* £0·97

3 £450, £450; £1085·70, £635·70; £1848·60, £762·90; £2500, £651·40

4 a £80; £20; £27·50 *b* 20% *5 a* £35; £8·20; £7·92 *b* 19%

Arithmetic—Answers to Chapter 2

Page 198 Exercise 1

1 382 000 kg; 700%

2 a television *b* 25% *c* (*1*) 135 hours (*2*) 360 hours

3 a 280 000; 310 000 *b* Ford

  c 20 000, $18\frac{2}{11}\%$; 10 000, $33\frac{1}{3}\%$; 20 000, $66\frac{2}{3}\%$

4 a 1950; 12 million *b* 1957; 7 million *c* about 10 million

5 b fifth day *c* second-third days *d* 5 *e* 54·4

6 b (*1*) seventh (*2*) fifth *c* yes *d* 7 weeks

Page 200 Exercise 2

2 b 15 *c* 2% *d* $\frac{7}{30}$ *3 b* $21\frac{1}{4}\%$ *c* $\frac{27}{80}; \frac{11}{20}$ *4 b* $\frac{3}{22}$

Page 202 Exercise 3

1 4, 4, 4 *2* 5, 5, 5 *3* 4·8, 5, 6 *4* 7·9, 8, 9

5 3·2, 3·5, 4 *6* £14·20, £14, £13

Page 204 Exercise 4

1 a 99·7 *b* (*1*) $\frac{1}{15}$ (*2*) $\frac{41}{75}$ *2* 57·4 cm *3 b* 15–19 *c* 20·2

4 a 11·3　*b* 11·2　　*5 a* 54·9 mm　*b* 54·9 mm　　*6 a* 70·0　*b* 69·8

7　£28·40 approximately

Page 208　Exercise 5B

1　61·5　　　　　　　　*2* £572·60　*3* 28·9 months

4　12 years 11 months　*5* 149 mm　*6* 57·5; 66·3

Page 208　Exercise 6

1 a 6, 7, 8　*b* 30, 33, 37　*c* 108, 113, 118　*d* 9, 11, 15

2　7 minutes; 4 minutes, 9·5 minutes

3　5·0 hours; 3·4 hours, 7·2 hours　*4*　0·51 litre; 0·50 litre, 0·53 litre

Page 210　Exercise 7 (approximate answers)

1　(i) 50, 37, 62　(ii) 50, 45, 56　*2 a* 61; 53, 71　*b* 22　*c* 56

3　£576; £549, £597　　　　*4 a* 29; 21, 35　*b* 28

5 a 12,10 years; 12,7 years, 13,2 years　*b* up to approximately 12,10 years

6　147 cm to 151 cm　*7 b* 56, 48, 67; 67, 62, 73　*c* 53, 66

Page 214　Exercise 8 (approximate answers)

1　30, 8　*2 a* 5, 1　*b* 11, 3·5　*c* 24, 5　*d* 9, 3

3　13 minutes, 2·75 minutes　*4* 7·1 hours, 1·9 hours

5　0·05 litre, 0·015 litre　*6* £6·50, £0·80

7　1309 hours, 55 hours

8 a 60, 7; 63, 11　*b* 62·3, 63·1

9 a 4, 4·2, 3·6　*b* 8, 1·1

Page 220　Topic to Explore Exercise 1

1 a 8　*b* 11　*c* 15　*2 a* 12　*b* 30　*3* 2, 6, 12

4 a 4　*b* 12　*c* 24　*d* 24　*5* 30

Page 221　Topic to Explore Exercise 2

1　600　*2 a* 90　*b* 380　*c* 9900　*d* 1 001 000　*3* 2652

4 40·320   *5* 20   *6* 120

7 a 6   *b* 24   *c* 120   *d* 720   *8* 1000, 999   *9* 36, $\frac{1}{6}$

10a 2   *b* 4   *c* 8   *d* 2^n (*1*) $\dfrac{1}{2^n}$ (*2*) $\dfrac{1}{2^{n-1}}$ (*3*) $\dfrac{n}{2^n}$

11a 3   *b* 9   *c* 27

Arithmetic—Answers to Revision Exercises

Page 223   Revision Exercise 1

1   £262   *2* 55 500   *3* £36   *4* £212·63

5   £2·08; 13·7%   *6* £536·66   *7* 4–5 years   *8* £3856

9   £297   *10* £3240; £810 profit   *11* £1800; £76·50

12a £76   *b* £0·60   *c* £236·80   *d* £0·84

13   £0·84   *14a* £30·13   *b* £23·38   *c* £97·63

15a £249·72   *b* £25·60   *16* £9·24

Page 225   Revision Exercise 2

4   approximately 44·5 seconds   *5*   52·2, 57·2

6   A: 50, 7; B: 53, 11   *7*   99·8 kg, 1·1 kg

8 a 6%, 33%, 49%, 12%   *b* (*1*) $\frac{3}{4}$ $\frac{1}{16}$ *c* 48 g, 7·5 g

9 b 3·9, 2·4, 3·7   *10b* 1960: 2·34, 2, 1·7, 0·9; 1970: 1·48, 2, 0·9, 0·8

Arithmetic—Answers to Cumulative Revision Exercise

Page 229   Cumulative Revision Exercise

1 a (2)   *b* (1)   *c* neither   *2* £1370·50

3 a 24·3   *b* 0·7   *c* 15   *d* 0·95   *4* £8·96

5 a 23, 27, 31   *b* 25, 36, 49   *c* 24, 35, 48

6   £289   *7* $5·6 \times 10^6$ *8* 19%   *9 a* 1110   *b* 1000101

10   £8·50, £11·05, £12·75   *11* £20   *12* £450·50   *13* 3·8 cm

14a $\frac{1}{2}$ *b* $\frac{1}{6}$ *c* $\frac{1}{15}$ *15* $4·4 \times 10^2$

16a 14·7   *b* 0·617   *c* 0·114   *d* 25 100 cm³

17   1111110 cm², 1011010 cm³; 126 cm², 90 cm³   *18* 64

19   21 km; 78·6 cm   *20*   £56·25   *21*   0·076 mm

22a   4·2 cm   *b*   4·7 m   *23*   46·8 cm²   *24*   13 800 m², 471 m

25   653 m²   *26*   4·0 × 10⁴ km, 1·1 × 10¹² km³   *27*   128 cm

28   23 kg   *29*   354 dollars; £22·50   *30*   3·32 × 10⁹; 3·38 × 10⁹

31   67 km/h   *32*   £781·83   *33*   £44·49   *34a* by £1·80

35a   1·35   *b*   0·802   *c*   0·100   *d*   106   *e*   4·84   *f*   0·0450   *g*   24·2 m

   h   1·73 mm   *36*   £15·03   *37*   £83·30; £11·76

38a   £840   *b*   £248·28   *39a*   6·6   *b*   7   *c*   6·4, 1·2

40a $\frac{9}{25}$ *b* $\frac{12}{25}$ *c* $\frac{21}{25}$; $1 - \frac{21}{25} = \frac{4}{25}$ *41*   1, 6, 12, 8; 8, 24, 24, 8

42   1806

Trigonometry—Answers to Chapter 1
Note. Variations in numerical answers may arise because of the use of different methods of calculation.

Page 239   Exercise 1

1 a   (1·88, 0·68)   *b*   (2·50, 4·33)   *c*   (0·66, 0·76)

2 a   0·500   *b*   0·985   *c*   0·087   *d*   −0·866   *e*   −0·174   *f*   −0·996

3 a   (−0·68, 1·88)   *b*   (−3·28, 2·30)   *c*   (−0·16, 0·99)

4   (i) $(r \cos a°, r \sin a°)$ (ii) $(d \cos n°, d \sin n°)$

   (iii) $(c \cos A°, c \sin A°)$ (iv) $(c \cos A°, c \sin A°)$

5 a   9·06 m   *b*   4·23 m   *6 a*   30 km   *b*   23·64 km   *c*   18·48 km

7   (1)   5·20 cm right, 3 cm above   (2)   5·20 cm right, 3 cm below

   (3)   4·24 cm left, 4·24 cm below

8 a   (5, 36·9°)   *b*   (13, 67·4°)   *c*   (3·16, 18·4°)

   d   (2·83, 135°)   *e*   (10, 126·9°)

Page 240   Introductory examples

1 a   yes   *b*   3·42, 9·40

2 a   yes   *b*   no right angles in figure   *c*   yes

3   Draw BP perpendicular to the base to meet CA produced at P.

Page 242   Exercise 2

1   8·89   *2*   2·55   *3*   11·9   *4* $a = 6·14, b = 5·66$

5 $a = 11·3, b = 7·71$ *6*   25·7°   *7*   61·5° or 118·5°

8 a   34°   *b*   46°   *c*   5·78   *9*   5·46   *10*   25·2°

11 AC = 506 m, BC = 389 m   *12* DF = 10·6 km, EF = 8·95 km

13 PR = 12·2 km, QR = 10·8 km   *14* BD = 4·26 m, ∠BDC = 54·5°,

∠CBD = 65·5°, DC = 4·47 m   *15* BD = 908 m; height of hill = 521 m

Page 245 Exercise 3

5 $\frac{4}{5}, \frac{3}{4}$ *6* $-\frac{3}{5}, -\frac{4}{3}$ *7* $\frac{12}{13}, -\frac{12}{5}$ *8* 0·866

9 0·866   *10a* (1) $\frac{3}{5}$ (2) $\frac{3}{4}$ (3) $\frac{12}{13}$ (4) $\frac{5}{12}$ *b* (1) $\frac{56}{65}$ (2) $\frac{16}{63}$

Page 246 Exercise 3B

12a T *b* F *c* T *d* F *e* F *f* T *g* F *h* T *i* T

Page 247 Introductory examples

1 a yes *b* 10, 36·9°, 53·1°

2 a yes *b* no right angles in figure *c* no

Page 249 Exercise 4

1 4·36   *2* 4·53   *3* 6·12   *4* 6·19   *5* 1·48   *6* 54·4

7 a $a^2 = b^2 + c^2$, since cos 90° = 0 *b* (3) *c* (2)

8 5·59   *9* 3·16   *10* 2·69   *11* 24·5   *12* 34·6

13 158   *14* 14·2 km   *15* 21·3 km   *16* 5·50 km   *17* 10·5 km

Page 251 Exercise 5

1 $\cos B = \dfrac{c^2 + a^2 - b^2}{2ca}$; $\cos C = \dfrac{a^2 + b^2 - c^2}{2ab}$

2 $\cos P = \dfrac{PQ^2 + PR^2 - QR^2}{2PQ.PR}$; $\cos Z = \dfrac{ZX^2 + ZY^2 - XY^2}{2ZX.ZY}$;

$\cos D = \dfrac{DE^2 + DF^2 - EF^2}{2DE.DF}$

3 74·4°   *4* 29·6° or 29·7°   *5* 78·4° or 78·5°   *6* 112·4°

7 90°   *8* 44·9° or 45°   *9* 125·1°   *10a* W *b* V

11 no such angles exist; since $a + b < c$, no triangle exists.

12 92·4°, 87·6°

Page 253 Exercise 6

1 $\triangle = \frac{1}{2} ca \sin B$; $\triangle = \frac{1}{2} ab \sin C$

2 a 5 km² *b* 5 square light years *c* 9·74 m² *d* 98·5 km²

3 a   3·01 cm² *b*   43·1 cm² *c*   2·35 cm² *d*   12·9 cm² *e*   87·7 cm²

4 ·   5430 mm²   *5 a*   30 cm *b*   410 cm²   *6 a*   62·4 cm² *b*   374 cm²

7 a   238 cm² *b*   2·82 m² *c*   3·14r^2 cm²; the same to 3 significant figures

8   49° or 131°   *9*   3·08 cm or 7·82 cm   *10*   9·52 cm²

Page 254   Exercise 7

1 a   17 cm *b*   124·8° *c*   94·4 or 94·5 cm²   *2*   8·2 cm

3   3·03 cm; 4·86 cm²   *4*   BC = 5·68 m, ∠BAD = 71·8°, ∠ABD = 58·8°,
      ∠ADB = 49·4°, ∠BCD = ∠CBD = 73·5°

5 a   69·3 m *b*   18·4° *c*   86·6 m

6   786 km; 137·7°   *7*   11·1 km, 7 km

Trigonometry—Answers to Revision Exercise

Page 257   Revision Exercise 1

1 a   (3, 5·20) *b*   (2·60, 1·5) *c*   (0, 7) *d*   (−7·07, 7·07)

2   A(4·70, 1·71), B(0·35, 1·97); 4·35, 0·26

3 a   (7, 0°) *b*   (3, 90°) *c*   (2, 180°) *d*   (25, 73·7°)

5 a $\frac{5}{13}, \frac{12}{5}$ *b* $-\frac{5}{13}, -\frac{12}{5}$

6   (*1*)   40°   (*2*)   4·91 cm (*3*)   11·1 cm²

        (*4*)   41·4° (*5*)   8·71 cm (*6*)   6·60 cm

        (*7*)   39·3° (*8*)   18·1 cm (*9*)   101·6   (*10*)   74·6 or 105·4

7   9·42 cm, 6·31 cm   *8 a*   PR = 6·24 cm, QS = 15·6 cm *b*   46·8 cm²

9   94·5 cm²   *10*   24 900 cm²   *11*   377 km; 242·5°

12   194 m   *13*   41·8° or 138·2°   *14*   208 cm²; 66·4 cm

Trigonometry—Answers to Cumulative Revision Exercise

Page 259   Cumulative Revision Exercise

1 $\frac{4}{5}, \frac{3}{5}, \frac{3}{4}; -\frac{3}{\sqrt{18}}, -\frac{3}{\sqrt{18}}, 1$ *2*   (i)   4·69 cm (ii)   7·68 cm

3   (i)   33·9° (ii)   19·4° or 19·5°   *4*   579 m   *5*   173 m

6 a   90°, 16·3°, 25 km *b*   299° approx.   *7*   36·9°, 12 m

8 a (*1*) $\sqrt{2}$ cm (*2*) $\dfrac{1}{\sqrt{2}}, \dfrac{1}{\sqrt{2}}, 1$ *b* $-\dfrac{1}{\sqrt{2}}, \dfrac{1}{\sqrt{2}}, -1$

9 a (*1*) 1 cm, $\sqrt{3}$ cm (*2*) $\dfrac{\sqrt{3}}{2}, \dfrac{1}{2}, \dfrac{1}{\sqrt{3}}, \dfrac{1}{2}, \dfrac{\sqrt{3}}{2}, \sqrt{3}$

b $-\dfrac{\sqrt{3}}{2}, -\dfrac{1}{2}, -\dfrac{1}{\sqrt{3}}, \dfrac{1}{2}, -\dfrac{\sqrt{3}}{2}, -\sqrt{3}$

10 10·4 newtons, 6 newtons *11* 14·1 newtons, 14·1 newtons

12 25 newtons, 43·3 newtons

13a 0·358 *b* 0·857 *c* −0·656 *d* −0·629 *e* 0·875

14b, c, d are true *15b* (*1*) {0, 180, 360, 540, 720} (*2*) {90, 450}

16b (*1*) {−360, 0, 360} (*2*) {−180, 180} *17* 90, 45 *18* $x = 90$

19 *20* *21*

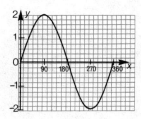
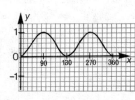
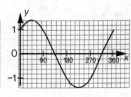

22 206 km west, 443 km north

23 A(1·29, 1·53), B(−8·39, 5·45), C(−4·70, −1·71), D(9·96, −0·87)

24a sin 80° *b* −cos 84° *c* sin 31° *d* −cos 10° *e* sin 55°

25 ∠A = 38·8°, ∠C = 90·2° *26* $e = f = 58·7$ *27* 7·5

28 452 square units *29* ∠Z = 90° *30* 130·5° or 130·6°

31 5·45 *32* 3·35 square units *33* 117 km, 122°

34 All are true except *f*. *35a* $\frac{4}{5}, \frac{3}{4}$ *b* $-\frac{4}{5}, -\frac{3}{4}$